Silmara Bernardino da Silva

Spatial logic, franchises, and language schools in medium-sized cities

Silmara Bernardino da Silva

Spatial logic, franchises, and language schools in medium-sized cities

The case of Presidente Prudente, São Paulo

ScienciaScripts

Imprint
Any brand names and product names mentioned in this book are subject to trademark, brand or patent protection and are trademarks or registered trademarks of their respective holders. The use of brand names, product names, common names, trade names, product descriptions etc. even without a particular marking in this work is in no way to be construed to mean that such names may be regarded as unrestricted in respect of trademark and brand protection legislation and could thus be used by anyone.

Cover image: www.ingimage.com

This book is a translation from the original published under ISBN 978-620-2-40088-6.

Publisher:
Sciencia Scripts
is a trademark of
Dodo Books Indian Ocean Ltd. and OmniScriptum S.R.L publishing group

120 High Road, East Finchley, London, N2 9ED, United Kingdom
Str. Armeneasca 28/1, office 1, Chisinau MD-2012, Republic of Moldova, Europe
Managing Directors: Ieva Konstantinova, Victoria Ursu
info@omniscriptum.com

Printed at: see last page
ISBN: 978-620-0-72090-0

Table of contents:

"Only once you've gotten to know the surface of things can you
that you can proceed to search for what lies beneath.
But the surface of things is inexhaustible..."

Italo Calvino

Chapter 1
1 INTRODUCTION

This book is the result of my work on my Bachelor's degree in Geography. The journey began when I first came into contact with the scientific research carried out by the Academic Research Group on Space Production and Regional Redefinitions (GAsPERR), based at the Faculty of Science and Technology (FCT) of the São Paulo State University (UNESP) *campus* in Presidente Prudente - SP. I joined the research group in the early years of my degree almost simultaneously with the approval and start of the activities of the thematic project entitled "Economic logics and contemporary spatial practices: medium-sized cities and consumption" funded by the São Paulo State Research Foundation (FAPESP), which is still in force.

During my time as a member of the research group, I worked as a Technical Support Fellow for the thematic project, carrying out activities linked to the operationalization of fieldwork, organizing *workshops*, meetings, transcribing interviews, as well as systematizing the data and information by categories and different themes that were being produced by the various research projects linked to the thematic project, from scientific initiation to post-doctoral research. The contact with a dense volume of data and information, its organization and initial treatment discussed with the professors linked to this major research and later the insertion of the materials into the Information Management Platform (PGI) was fundamental to understanding how this information can be obtained, how it should be worked with and how fundamental this methodological stage of the work represents in the overall work carried out.

Carrying out this task over the course of three years, together with the contact with the teacher-researchers and students, was fundamental to the emergence of a restlessness and desire to carry out research that covered some of the themes developed by the larger project.

The guiding objectives of this larger project are associated with the study of recent changes resulting from the restructuring of the city and urban areas, based on consumption. These transformations are based on three different analytical fronts: new logics for the location of companies, the spatial practices that reorient the process of urban structuring in its various scales and, as a result of the processes already mentioned, the restructuring of cities with new economic and social divisions of space (SPOSITO, 2011).

It was based on this larger project and contact with the various stages of doing research, as well as academic experience during undergraduate studies, that the intentions and ideas for building a research project that dialogued with at least one of the analytical fronts studied by the theme emerged. We therefore chose to study the *franchising* system for foreign language teaching in Presidente Prudente, seeking to understand the (re)production of urban space and the restructuring of the city with the insertion of this new type of business.

The research was therefore guided by the following objectives:

I. Identify the main dynamics that provide the spatial location of franchised and non-franchised foreign language schools;

II. Compare the differences between franchised and non-franchised schools and non-franchised schools located on shopping streets in the city center and those located in the suburbs;

III. Evaluate the location choices of franchised and non-franchised and polycentricity, according to the franchisor-franchisee relationship and independent owners.

3.1 Methodology

In order to carry out this work, we chose to work mainly with methodological instruments aimed at producing primary data and collecting secondary data, in an attempt to integrate quantitative and qualitative variables. The following sub-chapter is structured to describe these two work fronts, each with sub-items detailing the activities carried out at each stage. Fieldwork and interviews with the owners and/or people in charge of the schools were the paths taken to produce the primary data, , while the task of organizing the secondary data was to consult the *websites* of representative institutions in the sector, the *websites of* the brands themselves and public and private institutions, and a bibliographic survey.

3.2 Primary data

Primary data can be understood by us as data that has not yet been collected or organized, so the researcher is responsible for collecting and organizing it in order to obtain data and information that meets the specific needs of their research. Thus, one of the first activities carried out was fieldwork, after an initial survey on the *website of* the city's telephone company, seeking to locate the largest number of language schools in Presidente Prudente, whether or not they operated under the franchise system. Subsequently, a table was organized containing the list of schools and their respective addresses, which allowed us to go on *site to* check which schools were active, which had changed their location and which had closed. (See Table 1).

Table 1 Initial survey of establishments offering language teaching in Presidente Prudente - SP.

ESTABLISHMENTS	ADDRESS
EL SHADDAI IDIOMAS	R MAJOR FELICIO TARABAY, 100
CNA - LANGUAGE SCHOOL	AV CEL. JOSÉ SOARES MARCONDES, 1078
C.C.A.A. LANGUAGE SCHOOL	AV CEL. JOSÉ SOARES MARCONDES, 1005
BEST WAY	R SIQUEIRA CAMPOS. 180
PINK AND BLUE	R SIQUEIRA CAMPOS. 193
WESBRASIL- EXCHANGES AND COURSES ABROAD	DR. GURGEL, 940-A
CULTURA INGLESA - CENTRO DE ESTUDOS BRITANICOS	R BARAO DO RIO BRANCO, 657
PENNSYLVANIA - INTERNATIONAL EDUCATION	AV WASHINGTON LUIZ, 367
VIP TREINAMENTO E COMERCIO	R DR GURGEL, 277
MICROCAMP	R MAJOR FELICIO TARABAY, 821,827
UPTIME ENGLISH COMMUNICATION	R BARAO DO RIO BRANCO, 916
MADE IN THE U.S.A.	R PADRE JOAO GOETZ, 656- A
FISK SCHOOLS	AV WASHINGTON LUIZ, 525
CCAA - COLEGIO ANGLO AMERICANO	AV WASHINGTON LUIZ, 1350
PENNSYLVANIA LANGUAGE INSTITUTE	AV WASHINGTON LUIZ, 343
NORTH AMERICAN LANGUAGE SCHOOL	R DR GURGEL, 1014
YAZIGI - PENNSYLVANIA	AVDASAUDADE, 1.431
HISPANOAMERICAN LANGUAGE SCHOOL	R SANTOS DUMONT, 40
JP - HIGH SCHOOL, TECHNICAL COURSES AND COURSES	AV SALIM FARAH MALUF, 973, BLOCO I
OPTIMUM TECHNICAL LANGUAGE	AV JOÃO GONCALVEZ FOZ, 1789
CALENGLISH	R WALDEMAR FARIA MOTTA, 85
MASTER LANGUAGE INSTITUTE	R JOAQUIM NABUCO, 1200 -FUNDOS
SMARTZ SCHOOL	R TENENTE NICOLAU MAFFEI, 1120
FISK SCHOOL	R JULIO PRESTES, 1316
START ENGLISH	R ANGELO GALLI, 346
YAZIGI	AV WASHINGTON LUIZ, 981
WIZARD IDIOMAS	AV WASHINGTON LUIZ, 1064
LILI LINGUA CHINESA	R CASEMIRO DIAS, 277 - CASA 1
MAPLEBEAR CANADIAN SCHOOL	RALFANDEGA, 187
CNA - MORUMBI LANGUAGE SCHOOL	R PADRE JOAO GOETZ, 973
KUMON	R JOSE DIAS CINTRA, 219
KUMON	R COMENDADOR JOÃO PERETTI, 139 A
KUMON	AV CEL JOSE SOARES MARCONDES, 815
KUMON	AV REVERENDO CELSO ASSUMPCAO, 143
KUMON	R ANGELO MENEGUESSO, 263
KUMON	R ANGELO CALABRETTA, 81
KUMON	AV ANA JACINTA, 1969
KUMON (Regional Headquarters)	AV WHASINGTON LUIZ, 2431
KUMON	AV ONZE DE MAIO, 1011

Source: Lista Mais *website*, 2016.
Org.: Silmara Bernardino da Silva, 2016.

After organizing the table with the list of most of the schools that offer language teaching, fieldwork was carried out to check the current situation of each school.

situation of each school. In particular, those operating under the franchise system were also photographed, as can be seen in Table 2;

Table 2 - Photographic records of school facades of the schools

Source: Fieldwork, 2016
Org.: Silmara Bernardino da Silva (2016).

CNA Franchise

Franquía OPTIMUM TECHNICAL LANGUAGE

FISK Franchise

MAPLE BEAR Franchise

After surveying the schools, we identified which were franchises, the object of our analysis. We then contacted the nine franchised brands operating in the city by *email,* with a letter detailing the objectives of the research and the interview, to request interviews.

As the positive responses from the franchised companies to our request for interviews were insufficient, it was possible to diversify the profiles of the interviewees and our set of interviews was made up of the following profiles:

- 02 language school franchisees;
- 01 bilingual school franchisee;
- 02 local language school owners;
- 01 private language teacher;

The need to adapt the interviewees arose from the difficulty in agreeing to carry out interviews with franchisees, so semi-structured scripts were drawn up for the interviews with the various profiles of language school owners.

1.3 Secondary data

Secondary data can be understood as data already produced, organized and available for consultation from companies, public or private institutions, among other sources. In the research that resulted in this work, secondary data was essential for us to be able to complement the various facets of our object of study, because only with secondary information was it possible to organize and prepare the cartographic material that will help us in the construction of the text. After collecting the basic information such as the location, name and segment of the schools, we organized the data in a table and then searched for the geographical coordinates on *Google Earth* using the addresses, in order to proceed with the preparation of the cartographic material, as shown in Chart 3.

Chart 3 - Organizing the data on the franchises and their coordinates

School	CoordX	CoordY
FISK	'-51.388749	'-22.119088
Wizard	'-51.391205	'-22.118734
Yazigi	'-51.405544	'-22.137205
Yazigi	'-51.390939	'-22.119201
CNA	'-51.408930	'-22.129885
CNA	'-51.387780	'-22.121065
Optimum English Professional	'-51.409841	'-22.119000

Maple Bear	'-51.403509	'-22.117884
Kumon	'-51.391540	'-22.122227
Kumon	'-51.395592	'-22.117519
Kumon	'-51.386213	'-22.118937
Kumon	'-51.407679	'-22.119741
Kumon	'-51.400637	'-22.105810
Kumon	'-51.397766	'-22.135907
Kumon	'-51.430341	'-22.109540
kumon Principal	'-51.404120	'-22.115810
Kumon	'-51.397334	'-22.232882
Institute Brasi lei ro de Linguas -IBL	'-51.391485	'-22.120906
MICROLINS	'-51.383411	'-22.123604
MICROCAMP	'-51.386772	'-22.126185

Org.: Silmara Bernardino da Silva (2016).

This survey and organization made it possible to draw up maps showing the location of these schools superimposed on the urban fabric of Presidente Prudente-SP. Thus, with the maps identifying the location of the language franchises, we sought to find out the rules that regulate land use and occupation in Presidente Prudente - SP, which we found by consulting the City Hall website, was Complementary Law No. 153/2008 which provides for the Land Use and Occupation Zoning Law of the Urban Area of the Municipality of Presidente Prudente, together with Map 1, drawn up by the municipality's Planning Department.

Map 1- Zoning and land use in Presidente Prudente 2008.

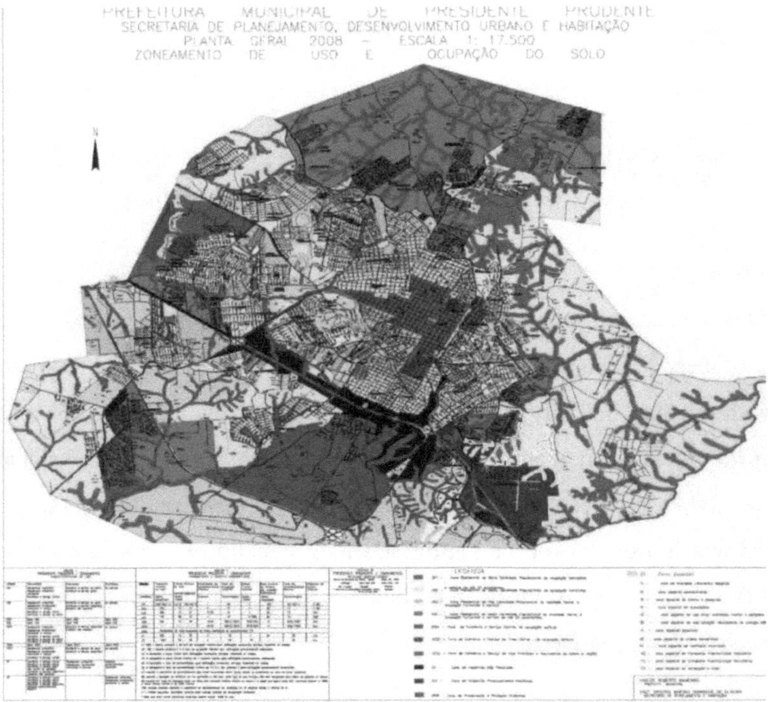

Source: Municipality of Presidente Prudente. Planning and Department.

On the official zoning map, we superimposed the map showing the location of the franchises in the city and drew up the cartography that allowed us to carry out analyses, crossing land use and occupation regulations with business strategies that have repercussions on changes in the structure of the urban space, as well as a comparative map between the location of the franchises and conventional schools.

Next, we selected the three largest language franchises in Brazil, which are present in the city studied. However, we thought it pertinent to present the number of units of the CNA and CCAA brands, which total 599 and 705 units respectively, so we can consider that these are the five brands with the largest number of units. The brands selected for the breakdown by station were: Fisk, Kumon and Wizard. We searched the *website of* each franchising company for the number of franchise units by state, resulting in Table 4.

Table 4 - Total number of FISK, KUMON and WIZARD units by federative unit

UF	FISK	KUMON	WIZARD
CA	1	1	2
AL	2	4	6
AP	2	2	4
AM	8	11	10
BA	31	35	29
EC	13	30	10
DF	9	134	29
ES	7	57	21
GO	25	30	34
MA	4	11	12
MS	15	36	12
MT	25	42	24
MG	52	151	118
PA	6	31	11
PB	10	10	7
PERNAMBUCO	7	15	19
PL	0	9	6
PR	60	128	99
RJ	64	88	111
RN	5	7	11
RO	8	8	5
RR	1	2	1
RS	28	40	112
SC	37	40	78
SE	3	2	8
SP	284	587	389
TO	6	3	3
TOATAL	705	1514	1181

Source: Consultation of the virtual pages of the brands: CCAA, CNA, Fisk, Kumon and Wizard.
Org.: Silmara Bernardino da Silva (2016)

After surveying the units throughout Brazil and organizing Table 4, it was possible to represent this information cartographically, as shown in Map 2 for the FISK franchise.

Map 2 - Fisk units by state.

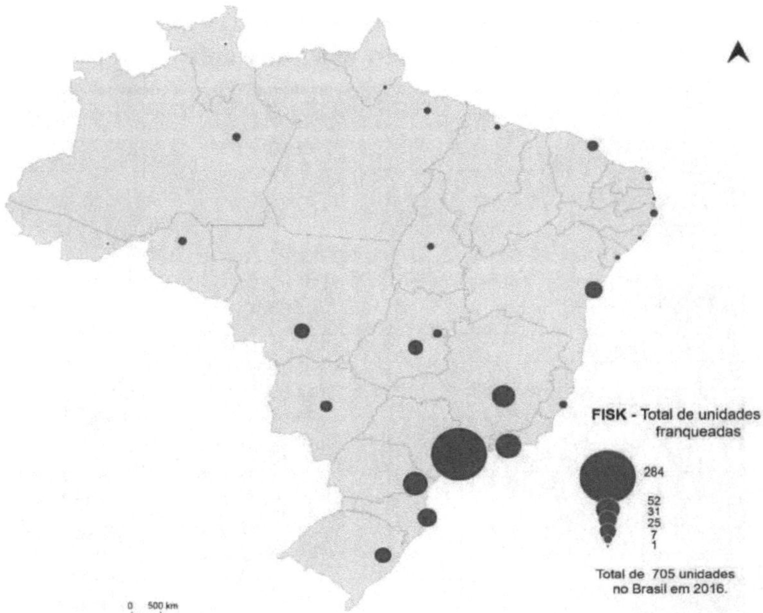

Source: Silmara Bernardino da Silva, 2016 (Prepared using *PHILCARTO software).*

After the quantitative survey for the three largest brands chosen to make up the count of franchised units, which allowed us to make the cartographic representations, we then decided to carry out a complementary survey, i.e. to detail the location of these franchises in each municipality of the Brazilian states, in order to classify each city based on the study "Region of Influence of Cities" (REGIC) carried out by the Brazilian Institute of Geography and Statistics (IBGE) in 2007 to classify the Brazilian urban network considering flows of information, goods and services. It is a hierarchical typology of centers according to their area of influence, taking into account the detachments of the population. After classifying the 1,016 municipalities in which the franchises are located, it was possible to draw up graphs that allow us to construct a comparative analysis of the three franchises that will be presented in the body of the text in order to support our reflections based on the relationship between the roles and functions played by the different urban centers.

1.4 Franchises

The interest in studying the location of language teaching franchises in medium-sized cities stems from the perception of the rapid changes in the structures of these urban spaces resulting, above all, from the globalizing phenomenon that has united different points around the world. The changes in the city are mainly related to the creation of new urban forms, such as commercial facilities and new housing arrangements, which influence the uneven expansion of the urban fabric. Influenced by new strategies, especially those linked to the reproduction of capital, which is transfigured by these arrangements and reflect the new economic orientation that connects all places to the dominant global logic in a different way. In the city, this is evidenced by the new location choices made by companies of different sizes and branches of activity. These changes are the result of transformations, above all economic ones, on a national/global scale, which have led national and (multi)international companies to adapt in such a way as to make it possible for them to reduce their profitability, In other words, the new logics linked to small and medium-sized businesses, especially local ones, which interact in everyday life, have contributed to making up the urban mosaic that is being formed. In other words, the new logics that companies adopt have led to processes that redesign and draw new intra-urban and, in some cases, inter-urban centralities.

With the density of economic changes, medium-sized cities have gained prominence in the urban network and play an important role in understanding the recent movement of urbanization, as the new centers of attention/investment of the agents that produce space.

Based on the reconfigurations in the roles of these cities, the larger thematic project, on which this study is based, selected six medium-sized cities for analysis - five from São Paulo, Marília, Presidente Prudente, Ribeirao Preto, Sao Carlos and Sao José do Rio Preto, and one from Paraná, Londrina - of which Presidente Prudente will be taken for the study on which this text is based.

With the transformations undertaken, above all, by the productive rearrangement with developments that respond mainly to the economic demands of accumulation and the constant adaptation of the market, the spatial dimension has reflected the impact of these changes. In this context, the franchising system was presented as the ideal object to guide the analysis, because, in summary, it contains the variables that are competing and complementary for the geographical analysis of the urban. The relevance is based on the fact that the franchise system presents itself as a possibility of adapting to the new forms of reorganization of capital in the face of the changes demanded by the market, especially in retail. Thus, the spatialization of franchises will allow us to grasp the new environments that are increasingly present in social life. Pointing to the tendency towards differentiation, complexification and extension of urban fabrics "and the complete redefinition of locational logics" (SPOSITO, 2009, p. 40), which the author presents as diffuse urbanization.

The franchise system can be read as a new way of operating the supply of products in the retail sector, which brings together international brands as well as those of local origin, which in the future expectation of expanding and maintaining their competitive capacity opt for this type of business.

The variety in the size of the companies that start operating under this system already demonstrates how the global, regional and local scales are connected and articulated in space, mainly due to the interdependent relationships between the various agents that contribute to carrying out franchise activities. In addition, the expansion of this type of company in medium-sized cities also highlights the transition of these areas, which were predominantly structured along productive lines, from the countryside, and which, due to the expansion of new information technologies, have contributed to shaping them as centers of consumption, since forms of travel and communication have become agile, confirming the importance of rethinking the dynamics and relevance of cities in the interurban division of labor corresponding to the scales of the network.

Studies dealing with franchises, their relationship with and influence on transformations in the urban space have recently begun to form part of a theme comprising multiple geographical research fronts, as this interest was predominant in other areas of the sciences such as administration, marketing, economics, etc, whose main objectives of analysis are aligned with the business perspective, i.e. they pay much more attention to the methods that allow economic agents to improve results, regulatory processes, types of transactions and accounting involved in the operating cycle, etc. (Ortigoza, 1996; Porto-Sales, 2014).

It is therefore up to geographical science to unravel the complex set of elements that have guided and are guiding the practices used by this group of agents, in order to transform the space, more specifically the changes resulting from the establishment of this type of business in medium-sized cities, as well as their inter-scalarity with other spaces.

In order to provide more detailed information on franchises, however, we need to present some elements relating to their origin and the mechanisms by which this type of business operates. The system can be identified by the following terms: *franchise, franchising* or franchising. The first term has its origins in France and is closely linked to the idea of concessions or exemption from monetary obligations to the feudal lord. *Franchise*, on the other hand, has more than one meaning and in a strict legal sense it means a right. "A right that someone acquires, to do something, by authorization or concession of another. " (CHERTO, M. 1988, p.1)

The same author defines a *franchise* as a method and arrangement for distributing goods and/or services. It is legally regulated through a contract made up of two parties: the franchisor, who owns a trademark, name or idea, method or techniques, secret, owner/manufacturer of a certain product or piece of equipment, *know-how,* who grants, i.e. enters into a contract with the franchisee, who is legally and economically independent of him, the right to exploit the set of elements mentioned above which make up his own system or part of it.

This requirement is reinforced by the support offered to the franchisee by the franchisor, such as the choice of location, installation, architectural design, courses on how to work with the sales system, stock control, advertising campaigns, etc? This varies greatly from company to company, and often implies higher or lower remuneration, franchise fee (pre-established and paid at the beginning of the contract), *royalties* (percentage on turnover, sometimes also on marketing) paid to the franchisor, who also plays a role in supervising / overseeing all stages of the business, which is characterized by Cherto (1988, p.6) as "*Business Format Franchising*" or in the author's translation as "commercial franchising".

In his thesis, Porto-Sales (2014) argues that the franchise system is the result of a spatial adjustment arising from the new configurations resulting from the productive and technological restructuring of the capitalist mode of production, which makes production and consumption more flexible and centralizes command activities in the distribution phase. The new organizational forms of work, the more centralized command and the new information and communication technologies (ICTs) have been reproduced in a dispersed way throughout the territories of production units and commercial establishments, forming new productive spatial circuits that we understand as the circulation of matter through the interaction of the different spaces of the total process, which are those of production, distribution, exchange and consumption of any product at a permanent pace.

(ANTAS, R. JR., 2012). Or in the author's words,

> [...] the franchise system was devised as a way of organizing the activities necessary to achieve consumption, characterized, on the one hand, by centralized command, expressed through the granting of know-how and monitoring; and on the other, by the dispersed territorial arrangement of points of sale, prescribed in a contract, designed to conquer markets at low cost and make a profit by stimulating consumption of brands. Spatial adjustment is defined by Harvey (2005) as the geographical expansion of the productivity factor, in which lies, illusorily, the solution to the contradiction of the circulation of capital, which derives from its tendency to overaccumulate and devalue. The geographical expansion of the product, the company, the workforce, the technique, and/or even the territory, was and has been conceived as one of the possibilities for reversing crises or even increasing profit margins and, in the case of the nation states, increasing their competitiveness on the international stage. (PORTO-SALES, 2014, p.65).

The analysis is aimed at understanding the transformations that this new way of organizing commercial activity has generated in urban space, restructuring and leading to the homogenization, which is never complete, of consumer practices that require and reproduce spatial practices.

In characterizing the current phase of transformation of urban space, in other words, diffuse urbanization, we need to point out that it was from the strategies to overcome the crisis of the capitalist mode of production that occurred in the mid-1970s that we reached the *status quo* of the consumer society, also defined as socially structured forms through which merchandise demarcates social relations and spatial relations (BARBOSA, 2004).

The subject of franchises and the choice of one of the various segments of this system stems from the expansion and importance that the numbers of this type of business have reached in the recent scenario of relative economic retraction. Graph 1 illustrates this by comparing the turnover figures for franchises in general with the gross domestic product (GDP) for the years 2008 to 2013.

Graph 1 - Percentage relationship (%) between evolution of the franchise system and Gross Domestic Product (GDP).

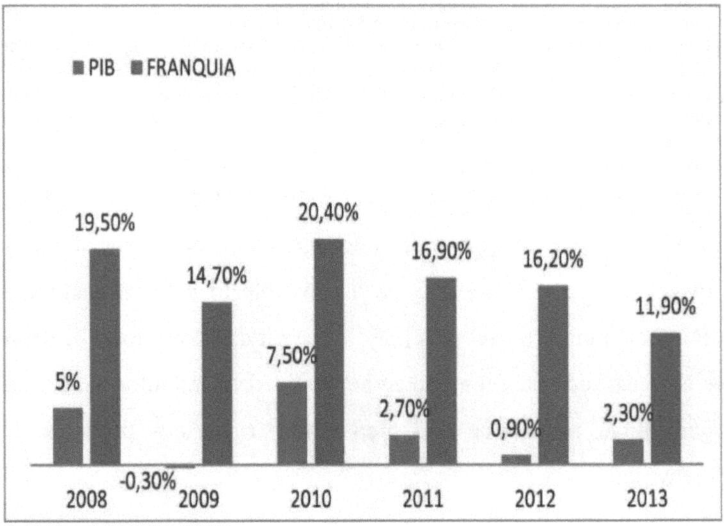

Source: ABF, IMF and RPEGN (2015)
Org.: Silmara Bernardino da Silva (2016).

The figures for the franchise system, despite having shown a contrast as observed in 2008 to 2009, with a drop of 4.8% between the periods, showed a more positive balance when compared to GDP, which showed a drop of 4.9%, taking its figure to a negative expression, reflecting the national scenario that showed lower or stagnant performance for the variables that make up the index. Although franchises have shown some slowdowns in the direction of lower numbers, we can say that the sector's performance in general, when we take the series shown in the graph, is quite favorable.

This brings us back to some of the guidelines that reinforce our aim of understanding how retail economic agents, franchisors, reinforce reflections on the transformations of the urban space. These characteristics are reinforced by the excerpt developed as part of the thematic project in which:

> Studies focusing on the spatial strategies of franchising companies are all the more important because, by taking as a basis the spatial structures already established both on the scale of the urban network and on that of the urban spaces where they are established, these companies redefine the processes of urban and city structure, as well as modifying the ways in which city dwellers use time and space.
> These changes do not occur homogeneously in all cities, but vary in intensity and form. There are significant differences between their presence in large, complex spatial structures such as metropolitan areas and those of medium-sized cities, where the entry of these logics and capitals can have an even stronger impact on established businesses and entrepreneurs, as well as on the ways in which consumers move around, choose places of consumption and redefine their use of everyday time as a result.
> The hypothesis that will guide the study is that the operations of these companies alter the inter-urban division of labor, reinforcing the tertiary roles of medium-sized cities and relatively diminishing those of small cities, such as Presidente Prudente, proposed here for the study, redefining the economic and social division of space, as they constitute vectors for accentuating the segmentation logics of consumption and consumers, from a spatial point of view.
> The effort to articulate the understanding of the economic dimension with the social dimension of the process under analysis is thus also a justification that we consider relevant for defending this proposal (SPOSITO, 2011).

In this way, we sought to draw up methodological procedures that would allow us to gather as much information as possible to help us analyze the spatial configuration that franchised schools play in the city. Below, we will detail how the procedures were carried out, mainly to gather all the data and information used, as well as the organization, processing and preparation of various products that helped construct the work.

Chapter 2

2 ECONOMIC-SPATIAL DYNAMICS AND THE TERRITORIAL LOCATION OF LANGUAGE SCHOOLS

Understanding contemporary urbanization requires the simultaneous articulation of contexts on the most diverse scales, i.e. we must consider the structures and conjunctures that articulate from the local to the global, without losing focus on the central scope of the research object.

To prepare this text, we have chosen the local scale, i.e. our object of investigation will be the city, which is where the multi-scalar process materializes, more specifically the medium-sized city, since the processes that lead cities to become more dispersed stand out, and through their forms and the processes that transform them will guide us to conduct the study along the path of understanding that these are instances of totality, according to Santos (2007).

> The study of totality leads to a choice of analytical categories that must reflect the real movement of totality. In addition to the categories of *time* and *scale*, which function externally, we must take into account the internal categories of *structure, function* and *form* (SANTOS, 2007, p. 199).

In this way, it is up to us to understand the city as a place that reflects in a particular way all the dimensions of contemporary life, both those linked to the collective and those linked to individual fulfillment, in other words, the complex social relationships that take shape in the present time after having an impact on form, This often leads to the process of intra-urban restructuring, without disregarding the political, economic and ideological influences that are often combined. In order to understand how these processes are taking place in Presidente Prudente, the following chapter looks at some aspects of the municipality's past formation.

2.1 Presidente Prudente, Historical Context and Middle City

We believe that the fraction used for analysis, the city, will help us develop a cohesive reading that takes into account the many external and internal dynamics involved in its constitution and allows us to see our object in its entirety. It is therefore necessary to present the main characteristics of our section, or more specifically, of the city of Presidente Prudente - SP.

Located in the west of the state of São Paulo, Presidente Prudente is the seat of the administrative region that bears its name, made up of 53 municipalities, and is located in the area of influence of the Tiete-Paraná Waterway. It has access roads that connect it to the capital of São Paulo, São Paulo and the state of Mato Grosso do Sul via the Raposo Tavares Highway (SP 270); via the Assis

Chateaubriand Highway (SP 425) connects with the state of Paraná and the western and northern regions of the state of São Paulo, and finally the Comandante Joao Ribeiro de Barros Highway, which connects several of the municipalities in the Administrative Region with the Paraná River. It also has a railroad operated by América Latina Logística (ALL) and Presidente Prudente airport. Map 3 shows the location and territorial extent of the municipality.

Map 3 - Geographical location - Presidente Prudente - SP

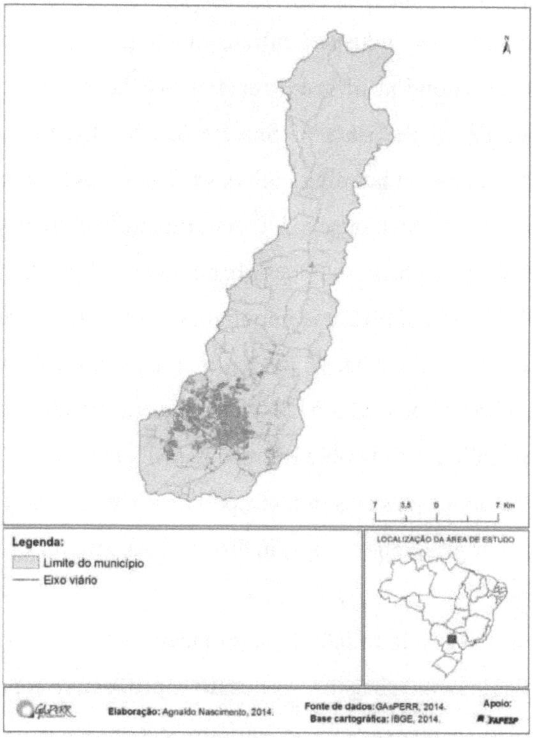

We will highlight some aspects of its socio-spatial formation that we consider important for understanding the current context of our study. The municipality's origins are strongly linked to coffee growing and the arrival of the railroad in the early 20th century, which were the main factors in shaping São Paulo's urban network. Since it was near the railway stations that the urban centers developed (Sposito, 2012).

Other activities, especially those linked to the primary sector, shaped economic cycles. O The development of livestock farming brought the first meat-packing plants to the city of Presidente Prudente, and in view of its extensive nature and the limited use of labor, it generated a population movement whose main reflection was the increase in the urban population to the detriment of the rural population. (Sposito, 2012)

This process took place during a period in Brazilian history in which priority was given to modernizing the countryside. The strategy adopted by the state for national development was to prioritize industrialization, which, geared towards the automobile industry, triggered the adoption of road transport as the main means of movement throughout the country. In the state of Sao Paulo, this led to the development of infrastructure projects aimed at building highways. The roads followed in the footsteps of the São Paulo railroads, which once led to the formation of urban centers, and which are now strengthening their position in the urban network (Sposito, 2012).

Still within this brief context, it is important to point out the repercussions of the 1970 World Crisis or the Oil Crisis, which led to a drop in profit rates, in addition to production costs and limitations, which led to the rearrangement of production patterns, shaping a new phase called the Flexible Regime, which consists of the deconcentration of the production sector of industries in metropolitan areas and their migration to more distant areas with lower production costs, in this case the interior of the state. (Sposito, 2012).

In this text, it is possible to cite the consequences that occurred in the cities of the interior of São Paulo, which began to concentrate productive activities coming from the capital, which led to investments in various sectors, especially those linked to commercial and service activities responsible for boosting the economy of the receiving cities. This adjustment in production did not lead to changes in the hierarchy of the urban network, since the metropolis maintained its influence in terms of power of command, concentrating the financial and management sectors of the industries, in other words, centralizing decision-making, since in most cases these transferred only the productive activities.

This set of transformations led to a significant change in the regional division of labor in the state of São Paulo, which became more significant in the mid-1990s.

> This dynamic was made possible by the increased possibilities for communication, both through improvements in infrastructure and means of transportation, making it easier and cheaper to move goods, and through the possibilities for circulating information, thanks to technical systems integrated by satellites, which allow a greater degree of connectivity between different points in the world (SPOSITO, 2001, p.193).

In addition to the aforementioned characteristics, material conditions at any time can be read through technology, since "technology is an element in explaining society and each of its geographical locations" (SANTOS, 2013, p.59). It is therefore coherent to think that the growing production of artificial nature, or of the technical-scientific-informational environment, converges with the (re)production of urban space (Santos, 2013).

This space is rationalized and (re)produced in a heterogeneous way, so it is possible to establish approximations between the different technological systems developed throughout history, we could speak of periodizations, since they allow us to unveil the changes and permanence, allowing us to perform the exercise of apprehending the differences between places, since the geographically unequal spatialization of technical systems can be explained by

> [...] the gaps between the variables that explain the differences in the organization of space [...] The technical basis of society and space is now a fundamental fact of historical explanation, since technology has invaded all aspects of human life. (SANTOS, 1994, p. 63).

The globalized technical systems are geographically unequal and the hegemonic logic that governs them coordinates the different economic, cultural and political functions characteristic of the globalization process (SOJA, 1993). As a result of the new global production order, the industrial deconcentration that took place in the state of São Paulo initially led to the strengthening of regional centers, governed by the logic of spatial organization, competitiveness and territorial specialization, with production as the structuring element of the urban hierarchy (SPOSITO, 2001, 2007).

We also have to consider the territorial division of labor, which shifted from hierarchical relationships to the juxtaposition of horizontal relationships, according to which medium-sized cities began to relate more strongly to other similar cities and also to larger ones. Nowadays, under the influence of the Third Industrial Revolution, land-use planning has become more fluid, with fewer obstacles imposed by the state. In addition to innovation, there is a strengthening of the niche of specialized tertiary activities, which are increasingly linked to a given territory.

2.2 Economic and spatial changes, medium-sized cities and franchises

The economic dynamics resulting from the concentration of capital, already summarized in the previous subchapter, have led to the growth of the consumer market in medium-sized cities and have also made them polarized by the spatial diffusion of new symbols, such as *shopping malls*, condominiums, super-hypermarkets, franchises, etc. Most of the time, these developments are coordinated by external agents linked to large national, international or multinational groups, or even by local

agents who work with external agents to keep their businesses afloat. Added to this are expectations that the new arrangements in the local company's *modus operandi will* become part of their profitability.

It is in this context that language franchises and the offer of this educational product/service draws our attention to analyze its elements in the context of a medium-sized city. We are interested in understanding how this expression, one of the many specialized materialities of the current capitalist mode of production, has reshaped centrality (ies) in the urban space.

We rely on Porto-Sales' (2014) understanding that centrality is intrinsically linked to the potential for location and that its explanatory elements are fundamental to the realization of the most diverse activities, and that they are aimed above all at productivity, which is understood as the efficiency that a given production factor performs "in increasing the profitability and/or profitability of the activity performed". (PORTOSALES, 2014, p. 30).

Map 4 is pertinent for elucidating our reflections so far, as it shows the actions of the different economic agents in the language teaching segment, i.e. we can see the presence of language franchises and non-franchised schools from other groups, as well as others of local origin. We can see that franchised and non-franchised schools are concentrated mainly in the central area and on the main thoroughfares.

Map 4 - Spatial distribution of franchises and language schools

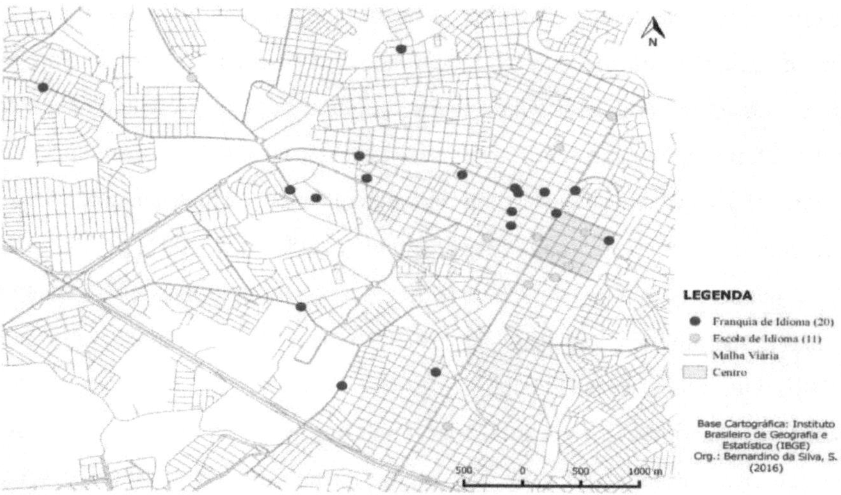

Source: Silmara Bernardino da Silva 2016.

The first aspect that the map raises for reflection is the quantitative relationship between language schools linked to the franchise system and non-franchise schools. The latter, usually local companies, account for ten units, while those associated with franchises total twenty units, i.e. the number of franchised schools is much higher, making up two thirds of the total in Presidente Prudente.

This reveals the territorial expansion of language franchises in the medium-sized city of Presidente Prudente, which is one of the elements that strengthens its role as a regional polarization in relation to the surrounding cities, in contrast to the decline in tertiary activities in these smaller cities. It demonstrates the correlation of strength, or competitiveness, that exists between local, regional and international capitals, because for the segment analyzed and in the surveys carried out for language franchises, it was possible to see the existence of groups whose power of command lies in other countries and even another continent. These dynamics have already been highlighted by SPOSITO (2007), addressing these trends in a more general way for franchises, and not specifically for language franchises.

As for the spatial location of these two types of establishments, we will first point out where they are located in the city. It is important to note that we are taking as our starting point and reference point for understanding the spatial location of the schools in the urban fabric, the center[1] , highlighted on the map by the polygon inserted in Map 4. The constitution of the center, which once represented the urban core that corresponded to the city as a whole in its founding period, has guided the urban structure and the subsequent and/or competing processes that have occurred in the past and recently have influenced the territorial growth of the city, which can unfold in other central areas (*subcenters, axes and shopping malls)* (SPOSITO, 1991).

In general, the map shows that the locational logic has been changing: considering the universe of language establishments in the center, schools not associated with franchises stand out.

In the central quadrilateral, there are three schools of this type, while the franchises are two units that, located within the central perimeter, are very close to Avenida Washington Luiz, which specializes in offering health services (consultations and exams), and some clothing establishments and other types of service and commerce, all aimed at the medium to high purchasing power segments.

In the area considered to be pericentral, which we considered to be located up to four blocks around the central quadrilateral, the opposite is true, i.e. there is a greater concentration of franchised schools, six units, while non-franchised schools total three units.

Within the context of changes in urbanization, we can see that franchising companies guide their franchisees to choose as locations for new schools central areas and sectors of the city that are more occupied or frequented by segments with greater purchasing power. On the other hand, in and around the city center, even if less intensely than in the past, we still find a significant concentration of other schools.

As Sposito (2001) warned, understanding changes in intra-urban space goes beyond simply pointing out the location, i.e. it is not enough to get to the essence of the processes that lead to its alteration. Therefore, before focusing on the reason(s) for the location of some franchised and non-franchised schools, it is pertinent to look at Map 5 to see the location and type of some of the new commercial forms, such as

---

[1] In the use of the expression "center" throughout the text, we want to emphasize that we are referring to the main center, or the original center, which was the one that shaped a monocentric city in the past.

shopping malls and *hyper-supermarkets* linked to large chains or groups.

The map shows five hypermarkets and two *shopping centers,* the area delimited as the main center is represented according to the urban zoning classification. We note that in the area bordering the center, located further south in the highlighted area, we find Parque Shopping, which has a unit of the Muffato hypermarket chain and the municipal bus station nearby. Prudenshopping

Prudenshopping also has a Carrefour hypermarket and is located very close to Parque do Povo, between Washington Luiz and Manoel Goulart Avenues, outside the central area. The Muffato and Walmart units are also located close to Prudenshopping. The Assaí wholesaler is located on Avenida Joaquim Constantino, in the south of the city, very close to the area of the city where the high-end condominiums are concentrated.

It's interesting to see how these new developments are linked to national and international groups and choose, since they have capital and the ability to influence the city's state and real estate agents, to be located far from the center. This demonstrates how the urban space has reshaped its structure based on different elements, globalization on the one hand and local agents on the other, but with actions that have a similar purpose, that is, the interest in obtaining greater income. It is in the face of this movement that their configurations are altered, mainly as a result of refunctionalizations and new urban creations.

Map 5 - Main commercial spaces - Presidente Prudente (2014)

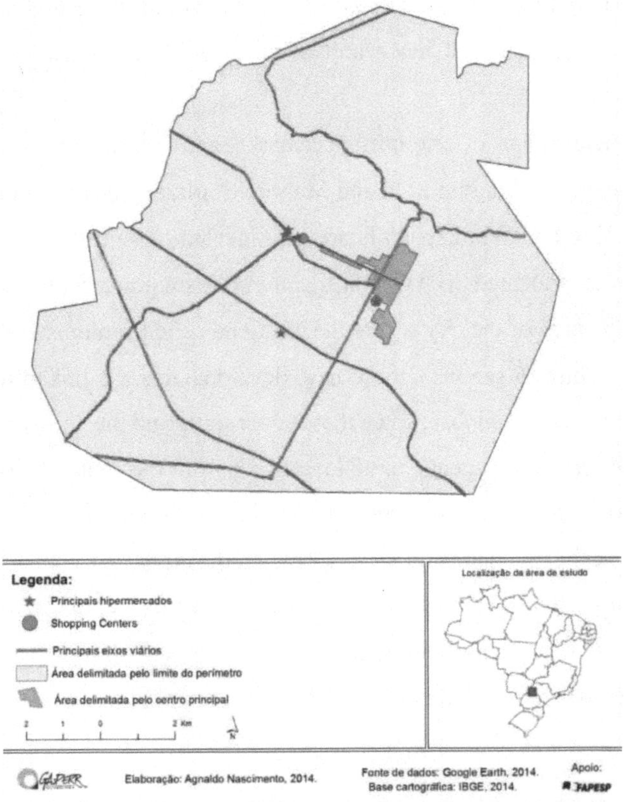

Presidente Prudente (SP). Main commercial spaces. 2014.

2.3 Presidente Prudente - Franchised and non Franchised Language Schools

Language schools were the forerunners of the franchise model in Brazil. The presence of 30 schools, franchised or not, reinforces the importance and consolidation of this type of business. We will present some characteristics of franchised schools such as MapleBear, Kumon, Instituto Brasileiro de Línguas, Wizard and Fisk. Even though these schools operate under the franchise system, each brand has its own specific characteristics (see Table 5).

Chart 5 - Characterization of some Language Franchises Franchises located in Presidente Prudente - SP.

	MAPLE BEAR I	KUMON	IBL	WIZARD	FISK
Origin of the brand	From 2005, in Vancouver, Canada - in Brazil the school is present in almost every state.	From 1558 Moriguctn, Japan brand is present on all continents In Brazil is distributed in all regions.	It was founded in 1996, but only in 2006 did it start operating under the franc hise franchise system. City of Mace".	The school was founded in 1986 in Campinas, but only started operating under the franchise system in 1987. It operates internationally, and in Brazil the school has units in in all all regions.	The brand was founded in 1958 in Sao Paulo, and it was only in 1962 that it began operating under the franchise franchise system. Its é international, and in Brazil it has units in in all states
Services offered	I Teaching offer I regular in I modality I bilingual	Mathematics, Portuguese, English and Japanese are offered. For the two first subjects, o can be offered as a kind of reinforcement course.	Offering of teaching the the following languages: English, Spanish, German, Japanese, French e Italian.	Offers courses in English, French, Spanish, German, Italian, Chinese, Japanese and Portuguese.	Offers courses in English, Spanish, IT e Portuguese for Brazilians
Methodology	Holistic holistic teaching in English and Portuguese, teaching focused on practical activities (experiments, arts, etc.).	Kumon is the name of the method applied by the school. It provides individual assistance, teaching materials self. It caters for all ages.	Methodology aims to speed up the learning process. Focus on conversation	Own methodology combined with the use of of technology.	Own methodology with a focus on the communicative approach via audio-oral.
Average monthly fees	From RS 1,500 to RS 3,000.00	On average...R$. 200.00	From RS 100.00 to RS 200.00.	On average.RS. 320.00	From RS180 to RS 200.
Initial Investment (RS)	From RS 100,000 to RS 750,000.	From RS 27,840 to RS 36.520	From RS 43,000 to RS 83,000.	From RS 50,000 to RS 125,000	From RS 83,500 to RS 498,000.

Source: Maple Bear, Kumon and Fisk company websites. IBLeWizarrt OrganizaçâozSilmaraBernanjino da Silva, 2016.

For the two non-franchised schools, Calenglish and Star English, we will present some information. Calenglish was inaugurated in 2011 and stands out for its classrooms, which allow for the simulation of various environments that could be experienced in everyday life. It offers English language courses. Start English began its activities in 2014 and also offers language courses in the English language modality.

The school's differential is its commitment to offering a language course at a lower cost than that offered by franchised schools, believing that the brand and the value of the course end up being an important factor in attracting mainly C and D class audiences. The interviews conducted with representatives of non-franchised language schools was an alternative due to the fact that the franchisees agreed in small numbers to grant interviews, as reported in the methodology. Conducting these interviews with non-franchisees allowed us to extract elements that enabled us to analyze certain aspects when compared to franchises.

It is interesting to note that the two non-franchised schools are located outside the perimeter of the center and its surroundings, and are close to Parque do Povo, an area that both interviewees considered to be an important point of centrality. This large green area is located in the southern quadrant of the city, where the most affluent segments live. The park and the two roads that flank it have an intense flow of both people and vehicles, the former for leisure and the latter because this axis connects to avenues that lead to regular and higher education establishments, *shopping centers* and hypermarkets. In addition, along Parque do Povo itself, there are commercial and service establishments, such as restaurants, bars, medical clinics, *pet shops*, etc., as well as public utility facilities.

Looking at the location of the schools, Calenglish and Start English are both located close to Parque do Povo, approximately 2.1 km apart, when we take the Quatorze de Setembro Avenue, one of the avenues that borders the park. Calenglish, despite being close to the park, is located in a commercial district, a kilometer or so away from the park and just to give us a reference, its proximity is to the east of the park and just to give us a reference to its proximity, its orientation is to the east of the park, closer to Av. Brasil. Start English is on a road that crosses the park, 30 meters from it.

When asked what led them to choose the location, both replied that the choice of location was justified mainly by the lower capital outlay for opening the school in terms of rental costs. The Calenglish school is owned by the interviewee's family, so there would be no need to pay rent, while Start English does pay rent, as the property is owned by the family of one of the school's partners, so the lease was more affordable.

Still on the subject of the business premises, both reported positive and negative aspects regarding its location. Start English pointed out the proximity to universities such as UNESP and UNOESTE, Parque do Povo and a snack bar, Apaí Raiz, as positive, because it considers that the mobility of vehicles and people in the area allows the school to be visible, and it also considers that its current location is an important factor in students opting for its language school. Regarding the possibility of other areas that could have influenced his choice of location, he was categorical in stating that when he was researching possible locations to open the school, the city center was his first choice, but due to the high price of rent he gave up; so the opportune proposal of a lower rent, as well as the ease of negotiating with the owner of his current building, was what weighed on his choice.

The interviewee from the Calenglish school, on the other hand, immediately pointed out the negative aspect of her business, claiming that she considers it to be very hidden, the street in which it is located has only two blocks, this is already a negative aspect, because it is very little busy, which makes visibility difficult, in addition to being difficult to find the location, even when searching on the *internet*, where there is an address with a street name very similar to that of the school, but which is incorrect. On the other hand, the positive points of the location concern firstly the space available, the school has a large building with two floors, a large parking lot and what caught our attention the most was her adjective: *"It's great because it's not in the center, because people... have an* ***aversion*** *to the center"* (Interviewee, Calengish School, 2016). Still according to the interviewee and her clients' perceptions, simply "crossing the center" is already a source of *stress.*

Asked which other areas of the city would be interesting for the business, she points out that the roads closest to Parque do Povo or Avenida da Saudade would be considered. With regard to this second option, we would like to point out that it is the location of one of the two YÁZIGI franchise units in Presidente Prudente.

We will now present the elements relating to location highlighted by the franchisees interviewed. There were three franchises we spoke to, and a brief description is in order, as one of them offers foreign language teaching in a different way. The franchisees interviewed are linked to the Instituto Brasileiro de Línguas (IBL), MAPLE BEAR and KUMON. The IBL brand can be characterized as a conventional language school, but offers the teaching of several languages, while KUMON offers courses aimed at teaching mathematics, Portuguese and English and Japanese language courses, a more diverse offer than other conventional language schools, while MAPLEBEAR is a regular bilingual school.

In order to get to know KUMON's spatial strategies, we interviewed a franchisee in the state of São Paulo, but in a different municipality to the one studied. This initiative was taken primarily because we considered it important to understand the brand's operating mechanisms, which have eight units in Presidente Prudente, as shown in Map 6, and having been unsuccessful in our attempts to interview one of them, we considered it important to bring information obtained from the interview, since one of the units in Presidente Prudente functions as a franchise school and KUMON's Regional Headquarters, so we can transpose general information about the brand to understand some of its operating mechanisms.

Map 6 illustrates the spatial representation of all the language franchises currently in Presidente Prudente - SP.

Map 6 - Spatialization of Language Franchises - Presidente Prudente (2016)

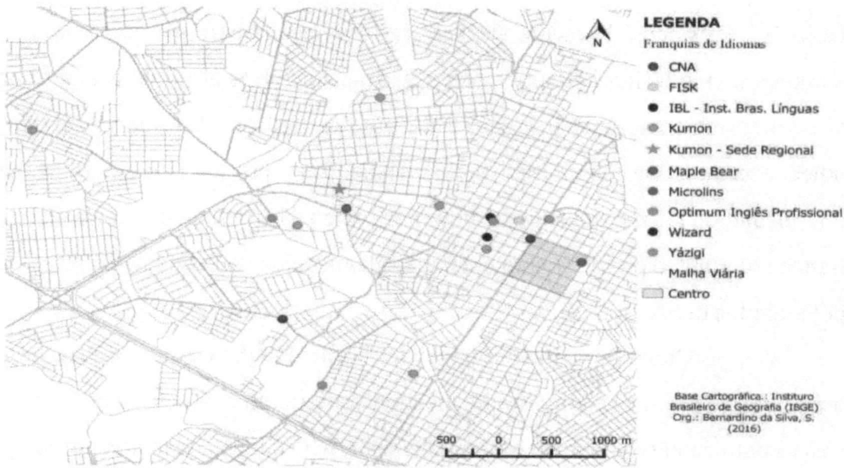

Source: Silmara Bernardino da Silva, 2016.

The Instituto Brasileiro de Línguas - IBL (Brazilian Language Institute), located in the city center, pointed out the easy access for students by public transport as a positive factor, as the center concentrates lines coming from all areas of the city. He also highlighted the structure offered by the city, specifically in terms of private parking. On the negative side, although it seems contradictory because of his previous statement as a positive point, it was the parking issue, i.e. the difficulty of finding parking because it is in the center, plus the fact that the road on which the school is located is one-way, which means that the negative aspect is related to the lower visibility of the school.

Some aspects of the franchisor-franchisee relationship with the Instituto Brasileiro de Línguas deserve to be highlighted, especially with regard to the choice of location. The city of Presidente Prudente came up as an option after the franchisee carried out a previous marketing study for the West of São Paulo, other regions of the state and other states. He pointed out that Presidente Prudente is a hub city and is the furthest away from other medium-sized or larger cities; in addition, the franchisee considered aspects such as the structure of the city, the economic power of the population, the degree of danger and the Human Development Index (HDI) and the GINI Index, which measures the level of inequality in cities. After analyzing all these variables, Presidente Prudente showed positive results and proved to be a potential market, making it the right city to open its franchise. It is also interesting to note that after choosing the brand, IBL, the franchisee carried out this study individually, in other words, it was an initiative that came from his own decision, without the intermediation of the franchisor, however, when he presented the chosen city, the franchisor granted permission to open the school, since it already had a study of future potential locations.

For the Maple Bear franchise, which is located in the vicinity of Parque do Povo, only positive aspects relating to the location were pointed out. The choice of location was a contractual requirement of the franchisor, who demanded that the franchised unit be located close to one or more main roads. This unit in Presidente Prudente is located very close to the Manoel Goulart and Quatorze de Maio roads. The proximity to Prudenshopping was also pointed out as an aspect that helps to form a positive opinion among the school's parents and students, since the target audience is children from nursery school to elementary school. In this way, the *mall* functions as a kind of "waiting room". The experience of the children's first day at the beginning of the school year illustrates this, as mothers drop their children off at school and go to the *mall* to, for example, have a coffee, walk around the *mall*, which would be a way of distracting them while they wait for an answer from the school about the adaptation period.

The school is next door to a Drogasil drugstore, which according to reports is a good thing, as a fair number of parents drop their children off at school and shop at the pharmacy. Being close to a police station and the UNIMED medical center also adds to the amenities described. What struck us in this interview was the emphasis given to the public square located in front of the school, as there is an appropriation of this public space by teachers and students in their daily educational activities, as well as the reference made to the calmness of the neighborhood, which, even though it is located near busy roads, *shopping centers* and the park, conveys a sense of security.

Look, I actually think it's a good place because it's quiet, isn't it? Despite this, we also have the police station here (nearby), there's the Unimed medical center which most of our children are affiliated with, and the *shopping center* helps because, for example, the mother has to wait for the child during the adjustment period. So she goes to the mall, walks around a bit, if she needs to we call, the pharmacy, sometimes the father leaves the child here and goes to the pharmacy, so in terms of location it's very quiet, it's a very peaceful neighborhood, we've never had a problem with robberies or anything like that, we haven't even heard of anything here.
First of all, they require us to be close to one or two main roads, but we have this one (Av. Goulart) and the other (Avenues around Parque do Povo) in our favor, even though there's a little river down here, we have this wonderful square, which we go to every now and then, so it's a good space. (INTERVIEWEE, MAPLE BEAR FRANCHISE, 2016).

When asked if there were any other areas of interest for the school in the city, the interviewee was the one who offered the most alternatives. The references were: a) the sub-center of the Jardim Bongiovani district, b) a corner plot of land opposite Prudenshopping that used to be a flower shop, and the rent was good, but the costs of adapting the area, renovating it, etc. were not interesting to the franchisee. did not appeal to the franchisee c) two spacious warehouses located near EUROMARKET, where a Korean car dealership used to operate, and which is considered to be a good location because it is served by roads, which would prevent parents from having to go through the main road with the heaviest traffic in the area. With regard to this third option, the interviewee said that the plan for this area is being studied by the franchisees, who are considering this possibility, as it is in the school's plans to expand its teaching offer, since from 2017 it will be working with Cycle II of primary education, which will make it impossible to maintain the current location due to the size of the school.

For the KUMON brand, what we will do is highlight some qualitative elements that we think are interesting to analyze when considering its location. In the interviewee's city, the school is located in a peri-central neighborhood, on an easily accessible street, between two regular schools, one public and one private, and is also close to the city's Tennis Club. The franchisor indicates a radius of installation for possible areas of interest, i.e. we understand that for cities in which the franchisor already has a franchised unit installed or those considered to be potential cities, it carries out a prior study of the ideal areas for installing new units, and keeps it reserved to indicate to potential new franchisees.

KUMON does not work with territorial exclusivity, and this is often one of the features offered by franchisors of other brands and is what attracts potential franchisees. By way of illustration, the interviewee's city has three units, each with a franchisee, and in Presidente Prudente there are eight KUMON units with different franchisees.

It is therefore up to the potential franchisee to choose their area within the radius indicated by the company and wait for approval. As far as the KUMON units located in Presidente Prudente are concerned, we can say that their distribution is relatively balanced, although they concentrate a greater number of units in a polygon that runs from Prudenshopping to the center, or in number we have the following distribution:

two units in

neighborhoods, three in neighborhoods, two on main roads and one next to a higher education institution.

After listing the positive and negative aspects considered by the owners of franchised and non-franchised language schools, let's analyze their spatial distribution so that we can make a comparison between these two types of educational services, based on the city's planning instrument, the Zoning and Land Use Law, drawn up and enforced by the municipal authorities.

In the previous chapter we presented Map 1, drawn up by the Municipal Planning Department, which shows all the areas of the city and their respective zoning, defined by the law "as the division of the urban area into zones, for which uses are defined (residential, commercial, service, industrial, institutional activities, etc.), and the rules for building on the urban plot (occupation)". (PRESIDENTE PRUDENTE. Complementary Law 153/2008). Based on the aforementioned cartographic reference, we drew up Map 7[2] overlaying the

---

[2] If you take a closer look at this product, you'll notice that there are some flaws. It's worth warning the reader that this was due to the difficulty in obtaining the cartographic base from the City Hall, since Map 1 is an urban plan drawn and processed using drawing *software* such as *AutoCAD*.

the franchised and non-franchised language schools on the zoned urban network.

Map 7 - Spatialization of language schools x franchises by zoning

Organized: Adapted from the Zoning Map drawn up by the Presidente Prudente Planning Department by Silmara Bernardino da Silva, 2016.

In continuity with the analysis we have been carrying out so far, we will now continue by mentioning the respective locations of the franchised and non-franchised schools and highlighting in which zoned areas they are located. We will start with more general observations, and then narrow down our observations to the language establishments of these two modalities, in relation to the establishments where it was possible to carry out interviews.

Of the total number of non-franchised (11) and franchised (20) schools surveyed, seven are located in the area highlighted in the map legend as ZCS1, identified as the Central Trade and Service Zone, which is vertically occupied and has its polygon filled in with brown, intended mainly for activities typical of the city's main center, characterized by the greatest possible diversity of retail, services, meeting points and social interaction, and is also an area of horizontal and vertical multi-family residences.

It is a high-density sector, both fixed (residents) and floating (consumers and visitors).

Based on the typology of uses observed in the highlighted law, the location of the franchised and non-franchised schools is consistent with the zoning proposal, since they have chosen to locate themselves in an area intended for commercial purposes, with a high intensity of flows, which we can read as the movement of people, vehicles, information, favoring consumption and encounters.

The areas highlighted by the blue color on the map are defined by zoning as High Density Residential Areas of Social Interest (ZR3), with horizontal and vertical occupations, with tolerated use, which is that which is not suitable for the area, but can be accepted in the zone in which it is located after consultation with the surrounding residents/owners and is regulated via documentation with the City Hall; for General Trade and Service, Specific Trade and Service. In these areas we find two non-franchised and two franchised language schools. As this is an area intended mainly for residential use, the existence of these commercial establishments fits in as a tolerated use. Article 20 of the same land use law, which deals with Residential Zones, states in its sole paragraph that "Other uses in residential zones should be considered as supporting or complementary accessories". Coincidentally, one of the schools is Calenglish, which, as mentioned above, had its location guided by family ownership of the physical space, and its dissatisfaction with the point is explained by the lack of visibility that is a possible consequence of its location, which has residential use as its main function.

There are also two franchised schools located in residential areas: KUMON and CNA. With regard to the CNA franchise school, its location is in what can be considered the subcenter of Jardim Bongiovani, close to UNOESTE, a large private university that operates in conjunction with the Regional Hospital, a reference point for city dwellers and residents of other cities for more complex consultations/treatments and exams. This sector of the city also has a variety of commercial establishments such as pharmacies, snack bars, clothing stores, medical equipment stores, as well as some franchise units in the food (Dasho), bar (Água Doce Cachaparia) and confectionery (Cía. do Brigadeiro) segments, confirming the thesis that franchises are looking to structure themselves in new central areas.

Only one school, Start English, is located in an area delimited as a Special Zone and classified as being of Sporting Interest. This characteristic is reinforced by the presence, above all, of public facilities designed to offer sports activities, such as the Presidente Prudente Municipal Sports Department (Semepp), which has ample infrastructure for practicing physical activities, as well as hosting the training space for some official teams; Parque do Povo is a public space that is also widely used for sports, and even though the Universidade Estadual Paulista (UNESP) is in a Special Interest Area for Teaching and Research, it has a Physical Education course, whose facilities are also used for sports activities, allowing for a mixture of uses. There are also some private spaces such as gyms, private and public sports courts, etc., which are located in areas designated by zoning for sports purposes.

This designation as an area of sporting interest was a positive aspect highlighted by the interviewee from Start English, who considered the mobility of people in the park to be very important, in terms of contact and visibility of the school. In this area, we also find the KUMON and Optimum Ingles Profissional franchises.

Yázigi is located in the Special Zone of the municipal cemetery, Avenida da Saudade.

We also have three KUMON franchise units and a school, among the yellow polygons that refer to the Medium Population Density Residential Zone (ZR2) with horizontal and vertical occupancy of up to two floors. The offer of educational activities also does not directly conflict with the main purpose, which is residential, as educational establishments are accepted as tolerated uses and complementary activities under the law that regulates the use of urban land.

2.4 Franchised and non-franchised schools: From Difference to Similarity

Having analyzed the spatial distribution of franchised and non-franchised schools on the urban network, we will now examine the differences in their forms and try to understand whether or not they are related to their geographical location. As a result of the fieldwork, we were able to obtain photographic records of the school façades, which will help us to describe the differences between the structures of the establishments. We'll begin this topic by looking at the physical structure of the non-franchised schools, Calenglish and Start English. Figures 1 and 2 show the structure of the schools.

Figure 1 - CALENGLISH School

Source: Fieldwork, 2016.

Figure 2 - START ENGLISH

Source: Fieldwork, 2016.

Before going on to describe the morphology of the schools in more detail, it's worth going back over the more general location aspects that we covered earlier and detailing more specific ones, such as the origin of the school, among others.

CALENGLISH, although located next to Parque do Povo, is in an area intended for residential use, which, as has already been pointed out, is an unsatisfactory location, mainly due to the low circulation of people in the area, even though it is located in a neighborhood considered to have a high population density. As can be seen in the image, the school has a regular structure, with a two-storey building, a small parking lot, but the surrounding area is considered easy to access/park in. It has a very striking façade, with the brand *layout* developed by a *marketing* professional who was also hired to take care of the school's advertising projects.

The origin of the school is closely linked to the interviewee's experience with the language. She took a language course, then did an exchange program and returned to Presidente Prudente in 2006. After returning, she decided that she would work as an English teacher, and in 2007 she had the opportunity to start teaching at , a Yázigi franchise school, which at the time only had one unit located on Av. Washington Luiz, and this contact with the franchise lasted a few months. After leaving the franchise, the interviewee, who lived in the Damha I condominium, started offering English classes in the condominium where she lived in 2008. As demand for the classes increased, they decided to expand the business by opening a school. The interviewee said she hadn't considered working with a franchise because she didn't have enough capital or a corporate investor at the time.

The school opened in 2011 and gradually grew, feeding off the payments received, since everything was invested in improving the school space. It can be seen that even though it is located in the neighborhood, there was concern about the architectural design, the advertising and the standard of the school, which has themed classrooms whose purpose is to simulate possible everyday situations that the students could experience in an international experience, in addition to the teaching materials, which are imported.

START English, which opened in 2014, is located on the central edge of Parque do Povo. It is also outside an area designated for commerce and services, and in a Special Zone of Sporting Interest, which for the owner doesn't affect the benefits of the location due to the large circulation of people in Parque do Povo and in the nearby snack bar, as well as the fact that its façade is positioned in the opposite direction to the traffic on Avenida Quatorze de Setembro, meaning that the intense flow of cars allows the school to be more visible. The non-franchised school has a very striking *layout*, and its internal spaces are similar to those of other schools in the sector, which we believe was influenced by the owner's previous experience in a franchised school.

Similar to the element that motivated the opening of the previous school, the choice to work in language teaching came about precisely because the owner took a language course as a student at the franchised CNA unit, and then went on to complete a university degree and postgraduate course in the language area. On completing the CNA course, he was invited to work at the unit and worked his way up through the ranks to unit coordinator. His desire to become an entrepreneur led him to think about and develop his own business project, as he felt he had sufficient knowledge.

Asked if he had considered working with language franchises before opening his school, his answer was yes and he stressed that he had the knowledge of all the stages from his experience of working in a franchised school. He had even looked for other franchise brands that were not yet present in Presidente Prudente, but according to him, they didn't create an identity either with the material he worked with or with the methodology offered, so he focused on developing his own system that would set him apart from the similar methodologies of the franchised schools.

There is a difference here in terms of the choice of location, since his first choice was the city center, which was only disregarded because of the high rents, although during the interview, when asked if he would choose another location for the school, he stated that he would remain at his current location, for the following reasons: decentralization linked to parking, traffic and the idea of "escaping the city center", which in his words "[....] makes it a little easier for the student to *get out of the center* and *be in* a *different place*." (emphasis added, Interviewee, Start English School, 2016) This statement is reiterated by him when at the end of the conversation I ask him about the reasons why students would choose his school and not a franchised school, he stresses that the location is one of its differentials.

After this more general overview of the franchise system, let's focus on the information and observations relating to those located in the city studied. The data presented refers to the brands Instituto Brasileiro de Línguas - IBL, KUMON and MAPLEBEAR. Figure 4 shows the façade of the bilingual education franchise in the city studied and the city of Sao Luís, MA.

Figure 3 - Maplebear Franchise - Presidente Prudente - SP

Source: Fieldwork, 2016.

Figure 4 - MAPLE BEAR Franchise - Sao Luís - MA.

Source: Fieldwork, 2016.

This brand is unique in that it is not a franchised language school that operates as a free course, since it offers regular bilingual teaching. With regard to its location, we can observe the convergence of two predetermined elements that are related by their nature and by the choice of location, one of which concerns the municipal zoning regulations that classify the area as a Central Trade and Service Zone, with vertical occupation (ZCS1), i.e. it has been designated for tertiary activities, with permitted use for trade and street service and tolerated use, which are those adapted to the area, for general and neighborhood trade and service. [3]

For MapleBear, there is a prior recommendation of possible areas and supervision by the franchisor, before the opening was approved, that the choice of location be contemplated by one or more of the city's main roads. This requirement was respected, as the school is located parallel to Avenida Manoel Goulart and its street is a continuation of Avenida Quatorze de Setembro. In addition to the requirements regarding the location, there are recommendations and inspections regarding the school's internal structure, such as the size of classrooms, safety equipment, and guidance on furniture. The franchisor recommends some stores located in the city of Sao Paulo to purchase furniture and objects that follow the standard set for its units.

It's interesting to highlight the franchisee's comments at this point, when she was asked about possible locations for the school, in which she reported immense difficulty in reconciling acceptance by the regulatory body, in order to obtain a license, and by the franchisor.

"Our biggest difficulty was the location, it took us almost two years to open. Because if the location was good enough for MapleBear, it wasn't good enough for the Health Department or the Fire Department. We had a lot of problems with the point, even a case down there with the Federal Police, we found a shack that would be great because it would be the way I wanted it, and someone came from São Paulo, a Canadian, he looked at it and authorized it, that was at the weekend, and we even invited him outside the rules to come and take a look, and he came and stayed here on Friday and Saturday and left on Sunday. And great, then the owner of the building told us he wasn't going to rent anymore because he'd forgotten that he'd promised it to a friend, that happened, wow that's absurd that you think about business, as we saw at the weekend we didn't sign the contract, we were going to sign it on Monday so the guy left. Then we had a friend of mine who has a lot of money, and she told me she was going to find me a place, so she found me a house on three plots of land, with lots of space, very nice, and her husband said to me, see a real estate agent and how much she values and what she values I'll close with you, coincidentally the real estate agent I got was their real estate agent, and she valued it at like R$ 4,000.00 the rent and he said fine. But then we called the fireman to look at it and the central corridor of the house was narrow, because it was an old house and he said it couldn't be done, because after the fire in Santa Maria, they had to have everything adapted, the corridor was very big and narrow and it couldn't be that size, and the support beam of the house couldn't be moved, gee, every time we fell, so we were almost giving up, and in the end we negotiated with the owner of this building (now a school) he has more than 1.000 properties in Prudente." He owns Euromarket, this hotel (IBIS) is his, that other one near the bus station is also his, he has a lot of things. He said to me, "Cícera, I'm not going to do business with you because I don't do business to lose, so I'm not going to do this with you," he said to me. Then we came to see him here, he was asking for R$15,000.00, it was an ugly, horrible shack, and he was asking for R$15,000.00, then he said look, I have a piece of land that I can build on for you, you know that Dashó over there on Bongiovani, have you ever been? Behind there they're building some towers, but there it used to be a farm and it was called Pé na Cova, so he said I have this land if you want I'll build it for you, so I said, but is there time to build it in

[3] Curiously, in the body of the law that typifies commerce and services, we didn't find any designation for franchised or non-franchised language schools; the closest we found were bookstores for vicinal commerce and services; publishing houses for general commerce and services; and schools, colleges and universities for specific commerce and services.

this timeframe? He did, so we agreed everything with him, he was going to build and by date, he could build for R$15,000.00 if he wanted to, but at least we'd have it ready and the way we wanted it.., then when they started the earthworks there, they found the brazilwood and Bugalho, who is a candidate today and is the Secretary of the Environment, was running for deputy or whatever, and whoever was in charge said that it would take too long and that he wouldn't be able to solve it here, so we lost the point again, and everyone said to us, you should have given up. Then in the end, when it came to the question of the brazilwood that there wouldn't be time to build it, he came to talk to us - so it didn't work out... - No, but you put a clause in there about a fine and we'd have to pay you R$50,000 if you gave up, I don't think it was R$30,000.00, we didn't give up, it's you who's giving up - Ah, but you found brazilwood - I said it's your problem, then he did the calculations, reduced the rent here and offered this sword for us to renovate (the current school site) but when we leave we have to remove everything we've done, walls, all that stuff, this place didn't even have (partitions). (Franchisee, MapleBear, 2016).

The excerpt from the interview already shows us a considerable interaction in the relationship between the agents who promote and produce the space, when we take franchises as a reference. The elements that permeate these interactions will be dealt with in more detail in Chapter 3, which deals with the (re)structuring of the city and multi-scalar relationships. For now, we present Villapa's (1998) reflections, which complement the situation experienced by the franchisee and her search for the ideal location on a theoretical level: "real estate development is the link between macroeconomic transformations and interurban restructuring".

In terms of architectural design, we can see a standard set by the franchising company when compared to other franchised schools in different cities. The bear logo and the leaf symbolizing the system's country of origin, Canada, stand out. There is even a standard for the classrooms, which must contain small children's toilets and tanks. The very arrangement of the furniture in the classrooms shows that the school does indeed have a predefined typology to follow.

The facade of the Instituto Brasileiro de Línguas (IBL) is shown in Figure 5.

Figure 5 - Instituto Brasileiro de Línguas franchise - Presidente Prudente - SP.

Source: Fieldwork, 2016

Figure 6 - Instituto Brasileiro de Línguas franchise - Americana - SP.

Source: Fieldwork, 2016

The Instituto Brasileiro de Línguas started its activities in the second half of 2016. As for its location, it also relates assertively to the urban zoning law.
urban zoning law, since it is located in an area designated for commerce and services, delimited by the acronym ZCS1, and very close to the central quadrilateral. The accessibility of being in the center was the decisive factor in its choice of location.
for its location. As for the external and internal physical structure (*layout) of* the space, the interviewee reported receiving support from the franchising company

both in terms of the orientation of the façade and the furniture, confirming the existence of a standard set by the franchisor, when we look at Figure 6, such as the franchised unit in the city of Americana.

franchised unit in the city of Americana, SP. A final aspect that emerged from the interview with the franchised school is the fact that it is the only school in the city that offers a variety of language courses for seven foreign languages, English, Spanish, Italian, French, German, Japanese and Portuguese, the latter aimed at foreigners living in the city.

We will now present some of the specific features of KUMON. For this franchise, there is very pertinent information on the location and physical structure of its units, since it has two types of establishments distributed throughout the urban fabric, and we reiterate that there are eight units in Presidente Prudente. The observations will be based more on those made in the field, since not all the information obtained from the interviews covered our spatial scope of study.

We will highlight our observations for this mark, characterizing the units by the areas defined by the zoning in which the units are located, sometimes more than one for the same zone.

The largest concentration of the brand's franchised schools is in the central commercial and service zoning area (ZCS1). This observation reinforces what was already highlighted when we compared the spatial relationship between non-franchised and franchised schools in all zoned urban areas. With the images of the respective units, we will see whether or not there is a typology in relation to architectural design recommendations, *layouts*, etc., as has been observed in relation to the other schools already analyzed.

Figure 7 Kumon

Figure 9 Kumon

Figure 8 Kumon

Figure 10 Kumon

Source: Fieldwork, 2016.

It can be seen that two of the four units located in the central area and situated exactly on two main axes, show notable differences between their commercial spaces. In Figure 7, located on Avenida Coronel José Soares Marcondes, considered an important avenue in the city as it has a varied concentration of commercial, banking and other services, there is a franchised school that is located in a residence, although it appears to be used for housing, but it also provides private educational services.

This configuration of KUMON franchised schools in residential units was a practice carried out by the franchisor in the past and which was changed more recently, without, however, excluding those that already carried out their activities in home units. The justification for this change can be found in the excerpt taken from the interview with the franchisee.

So today, in reality, this profile no longer exists. Because if you're going to franchise, be a Silmara franchisee, you can't anymore. In the past they used to say, be a Kumon tutor, you can do it in a room in your house, but that's no longer the case. *Nowadays, you have to have a commercial space* so that you can work with the children only, and so that you don't interfere with the parents, etc. *So whoever already has one, they're not going to close it if the franchisee is making a profit,* let's think from this point of view and sometimes, if they're going to build, for example, they need to see the standard and profile of the town they're in. (KUMON FRANCHISEE, 2016, emphasis added).
So that's why there's a lot of it in Brazil, but there's also a lot of it abroad in other countries... Japan is another reality, but I know several places in South America, Santiago, Buenos Aires... so it's a reality in Brazil, but it's different abroad. In Japan, in any cubicle, parents have children, for example, six or seven months old, who sit on the floor with the child to do... (KUMON FRANCHISEE, 2016).

On the other hand, the unit, identified in Figure 8, is located on Avenida Washington Luiz, which polarizes a considerable part of the city's traffic as it has shops and services aimed at the medium to high purchasing power segment. It is also a direct link between the central area and Prudenshopping.

On this same road, there are two franchised companies considered to be national brands in the language segment: Yázigi and Wizard. This leads us to think that they present themselves and establish themselves as direct competitors of the Japanese franchised school, above all because of their proximity to each other, but one detail that deserves special attention is the fact that this unit is also the KUMON Branch Office. This particularity is reinforced when we learn that this unit, or head office, is responsible for coordinating approximately 70 units of the franchise, which covers the Administrative Region[4] of Presidente Prudente, some cities belonging to the R.A of Marília, some cities in the region known as Alta Paulista and also some units located

[4] Cities in the Presidente Prudente Administrative Region: Adamantina, Alfredo Marcondes, Álvares Machado, Anhumas, Caiabu, Caiuá, Dracena, Emilianópolis, Estrela do Norte, Euclides da Cunha, Flora Rica, Flórida Paulista, Iepe, Indiana, Inúbia Paulista, Irapuru, Junqueirópolis, Lucélia, Marabá Paulista, Mariápolis, Martinópolis, Monte Castelo, Mirante do Paranapanema, Nantes, Narandiba, Nova Guataporanga, Osvaldo Cruz, Nantes, Ouro Verde, Pacaembu, Panorama, Paulicéia, Piquerobi, Pirapozinho, Pracinha, Presidente Bernardes, Presidente Epitácio, Presidente Prudente, Presidente Venceslau, Rancharia, Regente Feijó, Ribeirao dos Índios, Rosana, Sagres, Salmourao, Sandovalina, Santa Mercedes, Santo Anastácio, Santo Expedito, Sao Joao do Pau d'Alho, Taciba, Tarabaí, Teodoro Sampaio and Tupi Paulista.

in the state of Mato Grosso do Sul.

In this way, we can say that the centrality exercised by this KUMON unit goes beyond the scale of the city, as it also has repercussions at interurban and interstate level. Returning to the units located in ZCS1, it can be seen that the other units located on secondary roads already have the traditional structure of commercial spaces. This is probably because these units are more recent, as they already have the commercial pattern of exclusively private spaces, a change in requirement implemented by the franchisor approximately 10 years ago.

Figures 10 and 11 show units located on a territorial base defined by municipal regulations as a Residential Zone of High Population Density of Social Interest (ZR3) and point out, as already mentioned, that uses other than the original purpose, in this case identified as non-residential, are tolerated and classified as supporting or complementary accessories.

Figure11 Kumon

Figure12 Kumon

However, the difference between their spaces located in the same areas of definition, implies the following observations in Figure 11, the unit is again residential, with a small sign indicating that there is a franchised Kumon unit, present in a secondary neighborhood road, we understand that the circulation of people and vehicles should probably be of moderate to low intensity, and we suppose that this circulation, most of the time, is carried out by the residents of the neighborhood and its surroundings.

As for Figure 10, its physical structure is more similar to that of commercial establishments. First of all, it's important to note that the current location of this unit is recent, and probably took place between 2015-2016, as a search on the franchisor's *website* found the address of Avenida Onze de Maio, one of the avenues that surrounds Parque do Povo, connecting the northwest and southeast areas of the city. Even though its location has changed, from a main road and a Special Sports Interest Zone, to a secondary road in ZR3, we understand that this may have had less of an impact, due to the fact that the current address is on a parallel road with easy access to Avenida Onze de Maio. There are some commercial establishments located on the same road that contribute to making the area more dynamic, including the Guaíba barbecue restaurant, Massa Pura bakery, post office, some clothing stores, etc.

Figure 13, located on Avenida Ana Jacinta, which is a sub-centre, has this characteristic reinforced by the zoning that classifies it as a Trade and Service Zone for Roads, of vertical occupation, with the same permissions and tolerances applied to the ZCS1 already highlighted. Thus, we can consider that its location is in line with the purpose established by the official body, since being located in a subcenter is a positive factor for its visibility due to the greater number of individuals who circulate through this small business, as well as the fact that there is no competition for the service offered in this space, since it is the only establishment that offers language teaching.

For Figure 14, the classification is Residential Area of Medium Density Population, of horizontal occupation (ZR2), although there is no prohibition as to the commercial and service activities that are allowed and tolerated, it presents characteristics that its activities are essentially ancillary to the neighborhood, mainly because its structure is residential and its advertising mechanism, a small banner hanging on the wall where it is possible to see the brand name and the contact telephone number. The road on which the unit is located is of secondary importance and probably repeats the opposite mechanisms to the large circulation of people and vehicles, leading one to consider that its public or radius of reach is that of the neighborhood itself.

Figure13 Kumon

Figure 14 - Kumon

Based on the characterization carried out for both franchised and non-franchised schools, we can point to elements that make them similar, especially the aspects relating to physical structures, as both were concerned with developing architectural projects that reflected, above all, the school's proposal and purpose, and there was also a similar development of advertising projects that allowed for greater visibility of their brands. As far as the franchises are concerned, it is clear that the franchising company provides guidance on the characteristics mentioned for the three brands interviewed, and even for other franchises in the city, a simple search on their *websites* is a positive aspect highlighted by the company as a means of influencing the choice of a potential franchisee. There are also similar mechanisms for franchises in terms of greater flexibility when it comes to choosing a location, considering the new areas and even alternative points in relation to the city center. This possibility is more limited for non-franchised schools.

We can therefore say that there are common strategies for offering language services between franchised and non-franchised schools, and different strategies between them, as well as similar and different strategies between the franchised schools themselves.

Chapter 3

3 MULTI(POLY)CENTRALITIES: A LOOK AT LANGUAGE FRANCHISES

In this chapter we will present how language franchises and their locational choices have influenced transformations in the structure of urban space, as well as changes in their role in the urban network, based on the multi-scalar processes that have taken shape.

Explaining the recent changes in urban space implies considering the dynamics and processes resulting from the social spatiality of this space, which, in the differentiation and integration of its areas, conforms to the totality of the urban. "As a reflection, medium and social condition, human spatiality is the result of spatial processes, creators of forms and spatial interactions" (CORRÉA, 2016, p.128).

Before delving into the multiple processes that shape the urban reality studied, it is necessary to emphasize our agreement with the idea put forward by Sposito (2007), who understands and develops the use of the term "average city" as a possible methodological way of unveiling the changes that urban spaces undergo. The term is still used in many studies only in terms of its demographic dimension, which would fail to grasp the complex changes in the relational profiles between different cities. Studies of the urban network were based on a rigid foundation structured through urban hierarchy, and this past characteristic is no longer capable of explaining urban contemporaneity on its own.

The author also outlines the processes to be observed when carrying out studies of medium-sized cities.

> What, then, is new for the study of medium-sized cities in terms of the relationships between the intra-urban and inter-urban scales?
> In addition to the hierarchical and horizontal relationships that were established in previous technical periods, there are relationships that we will characterize as transversal. They can be called such because they go beyond the network in which they are inserted [...] (SPOSITO, 2007, p.237).

These premises underpin and guide our analysis of the medium-sized city of

Presidente Prudente.

3.1 Location and Urban Structuring: The scale of the Intra-urban Space

The city can be understood as originating from spatial processes that result from the force of attraction, cohesion or as a function of agglomeration economies that result in spatial concentration (Corrêa, 2016).

Investigating the city's production process requires us to start from the consolidation of its main center and an understanding of its role, since it is from its structure that the territorial expansion of its urban fabric occurs. It is in the central core that the conditions for the reproduction of the workforce are traditionally found, as well as greater flows of circulation and supply of goods and production of services, in other words, it is and has been the main reference point for urban centrality and symbolic centrality.

Urban territorial expansion leads to the structuring of the city as a whole, which is shaped like a mosaic, in which each part differs from the other in size, color and shape, but which together constitute a unit. In this way, the city as thought concentrates a range of material diversities, geographical and immaterial objects. And the relationships between the different social groups are similar to the parts of the mosaic due to the different uses and appropriations they make of urban land, which when interconnected make up the urban mosaic.

Differentiating land use on the basis of urban activities and functions in an attempt to relate them to the new spatial configurations means that we need to point out one of the original elements in its formation process. The aspect we want to highlight here is the action of the real estate agents who command and guide the dynamics of the structuring of space. Since most parts of the territory were and are privately owned, the owners of these areas, by offering them on the market, create expectations, on the one hand for the owners who make a profit by renting or selling them. On the other hand, they create expectations for those who are competing for a certain area in order to overlap it with one of the many specialized activities (SINGER, 1982).

It is in this way that the urban fabric takes on new configurations and uses, in other words, the coexistence of economic (industries, commerce), residential (houses, housing developments, condominiums) and leisure (public facilities) activities materializes the city as a place of consumption and production, simultaneously. This is where the productive and reproductive forces exchange through circulation (SPOSITO, 1991).

As a producer and product in constant movement, urban society is made up of various subjects who organize themselves into classes based on their political, economic, social, cultural, etc. unity, Thus, in general, we can speak of social agents who produce urban space, such as the owners of the means of production, landowners, real estate developers, the state and excluded social groups (Corrêa 1995), who act on space according to their affiliations, and by producing parts of the city in their own way, their consumption will also occur in different ways.

The structure of the city must be thought of from the point of view of the various uses and agents, bringing elements of urban concentration and dispersion into its understanding.

Representing spatial concentration is the city center, which from its past formation provides us with important clues to unravel current changes in its role and relevance in the urban network. The evolution of transport and technical systems has allowed the urban network to expand into a larger city, leading to the emergence of new commercial and service areas, or new central areas (Sposito, 2013).

The emergence of these new commercial areas, together with new types of human settlement, have led to the emergence of new spaces in the city that generate centrality, such as main axes, sub-centers, new consumer equipment, which, by concentrating in a new area of the city, will be shaped by specific social parameters, creating spatial differentiation.

We saw this when we looked at the spatialization of franchised and non-franchised language schools. For the non-franchised schools, we noticed that the locational pattern predominates in the area of the main center and a few distributed throughout the neighborhood or sub-center, reinforcing the already existing dynamics of centrality, while for the franchised schools we observed the opposite situation when we saw that they are located in a more dispersed way throughout the fabric, almost accompanying the new commercial areas. Few franchised schools are located in the center, but their distribution is more balanced, with schools located on the main axes that meet the consumption demands of the various social segments, schools in subcenters and in the neighborhoods themselves.

Reflecting the recent movements that have transformed non-metropolitan urban spaces, the creation of new central areas is part of a process that takes place over the long and medium term, and consists of the transition of the city's central spaces and the dynamics that accompany it. Previously, the predominant structure was that of a monocentric city, i.e. its main center was the area where the main stores, services, bank branches and official institutions met, concentrating in its radius the attraction at a high level of quantity, density and origin of the flows of people, information and finance, these as reaffirmers of its centrality. As a result of transformations in the ways in which people are deployed and in technical systems, new areas have been created which have concentrated the supply of goods and services in different areas of the city, leading to the creation of central areas which generate centrality. These new spatial arrangements, such as sub-centres, peri-central regions and main thoroughfares, have created a new centrality, which can also be read as a relative reduction in the capacity of the traditional center to polarize flows, configuring multicentrality (Sposito, 2013).

More recently, the relationship between central areas, the resulting urban morphology and centrality have taken on a new shape, although this does not exclude the conditioning factors of past central structures; on the contrary, they have come to coexist. Polycentrality is identified as the emergence of large commercial and service areas, imposing new spatial structures that are different from those that have existed up to now.

The differentiating element is the degree of attraction of the polycentricity that has polarized the city as a whole, as opposed to what had been the case up until now.

to what occurred in spaces characterized by multicentrality. As for the centrality exercised, for the polycentric central spaces, it is no less than that developed by the main center when considering the range of goods and services on offer or the level of specialization. The motivating elements for this are not the sum of local supports, or horizontal relationships that are progressively established, but their external or vertical supports that selectively distribute their products throughout the space, reinforcing socio-spatial segmentation and segregation (Sposito, 2013).

In this way, we want to emphasize the guiding elements of locational choices. Previously, new areas were built in continuity with the main areas of the city with the greatest circulation of people, mainly due to accessibility. Today, the influence of the real estate aspect is a fundamental driver in the configuration of new commercial areas. It is hoped that empty areas will allow for the creation of ample commercial space, enabling both commercial profit and profit from future land appreciation. Concern about providing an accessible urban area does not fit into this situation. The excerpt below summarizes our thoughts.

> [...] it is necessary to evaluate this decentralization, because it does not reveal the dispersion or distribution of traditionally central activities throughout the city, but on the contrary, it reveals new forms of centrality. By denying the concept of a single, monopolizing center, it recreates centrality, multiplying it through the production of new structures that allow for new forms of monopolies, because they (re)specialize and (re)specialize commercial and service activities, reproducing central conditions and qualities in other areas of the city.(SPOSITO, 1991, p.13)
>
> The new choices, those made by the big developers responsible for setting up the new commercial and service structures, do not seek out the city, but rather seek it out and provide support for the city and its inhabitants to seek it out (SPOSITO, 2013, p.77-78).

The urban structure that predominated until then has been overcome, in other words, the center-periphery logic is overcome by conditions that reorient locational choices, leading to diversification. This change makes up the scenario of urban restructuring, leading to the formation of a poly(multi)centric city that interacts in different ways with its residents, Those who have their own means of transportation consume the city and the city in different ways or in its entirety because they have autonomy of choice in their journeys, while those who depend on public transportation only consume in the areas where they move around, or their circulation is much more limited to the mono-multicentric areas of the city (Sposito, 2013).

3.2 Interscalarity, Medium-sized Cities and Franchising

The complexification of the urban network through the structuring of a more open city involves interactions with cities in the network in which it is contained, as well as with other networks in the country as a whole and sometimes even with locations in other countries. This condition is synthesized by interactions that result from the following dynamics:

> the differentiation of the functions and roles of cities and the creation of hierarchical patterns of different typologies and the movement of the capitalist mode of production that spreads across territories, processes such as industrialization, globalization and the financialization of economic relations, building, according to Santos (2001, 2008), a corporate urbanization, the result of the territorialization of companies in the world (CATELAN, 2013, p.65).

We found this to be the case when we looked at our object of study, franchised language schools. The existence of nine brands in this segment and the number of franchised units located in Presidente Prudente reinforces the fact that urbanization today is characterized by corporate use, which makes urban space the general condition for capitalist (re)production. For the franchise system, this is evident when we present data from the sector in a survey carried out by the Brazilian *Franchising* Association (ABF)[5] in 2016, which shows the total number of franchised units in relation to the total number of municipalities. Of the 5,570 cities in the country, 2,321 have some form of franchise, representing 42%.

Having said that, we chose the FISK, KUMON and WIZARD brands to carry out a more detailed survey that would allow us to understand their spatial distribution in country. Maps 8, 9 and 10 represent the spatial distribution of these three brands, which, in a survey published in 2016, were among the 50 largest franchise brands in all segments.

---

[5] Available at: .www.abf.com.br/numeros-do-franchising/

Map 8 - FISK franchises by state.

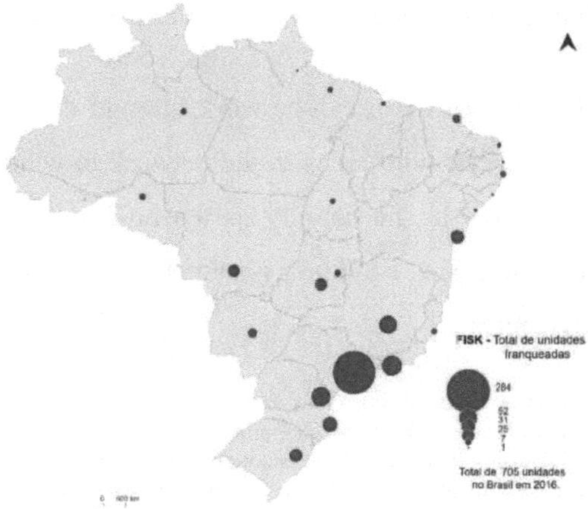

Source: FISK company *website.*
Elaboration: Silmara Bernardino da Silva, 2016 (Prepared using *PHILCARTO software).*

Table 6 - Distribution of FISK units

UF	№ UNITS	UF	№ UNITS	UF	№ UNITS	UF	№ UNITS
CA	1	GO	25	PE	7	SP	284
AL	2	MA	4	PI	0	SE	3
AP	2	MS	15	RJ	64	TO	6
AM	8	MT	25	RN	5		
BA	31	MG	52	RS	28		
EC	13	PA	6	RO	8		
DF	9	PB	2	RR	1		
ES	7	PR	60	SC	37		

Source: Consultation of the KUMON franchisor's *website*.
Org.: Silmara Bernardino da Silva (2016).

Map 9 - KUMON franchises by state. federation.

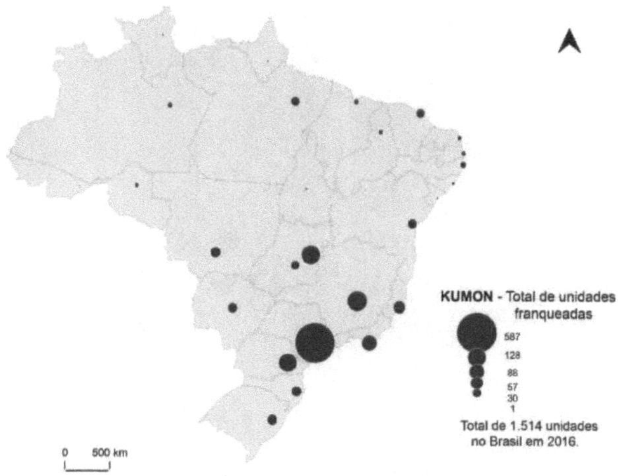

Source: Consultation of the KUMON franchisor's *website*.
Organized: Silmara Bernardino da Silva, 2016 (Prepared using *PHILCARTO software)*.

Table 7 - Distribution of KUMON units

UF	№ UNITS	UF	№ UNITS	UF	№ UNITS	UF	№ UNITS
CA	1	GO	30	PE	15	SP	587
AL	4	MA	11	PI	9	SE	2
AP	2	MS	36	RJ	88	TO	3
AM	11	MT	42	RN	7		
BA	35	MG	151	RS	40		
EC	30	PA	31	RO	8		
DF	134	PB	10	RR	2		
ES	57	PR	128	SC	40		

Source: Consultation of the KUMON franchisor's *website*.
Org.: Silmara Bernardino da Silva (2016).

Map 10 - WIZARD franchises by state. federation.

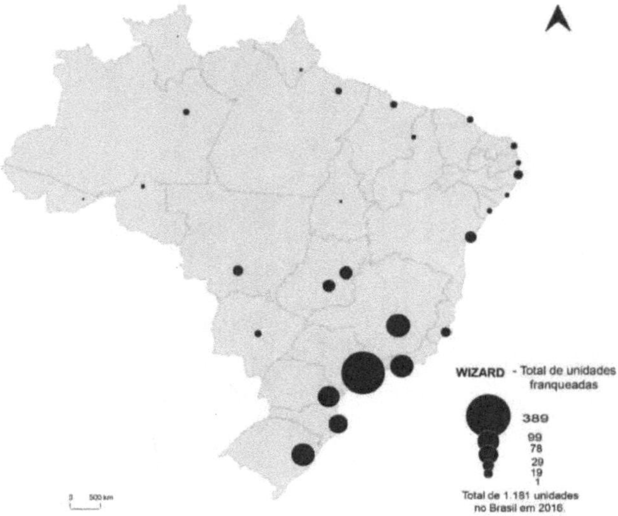

Source: Consultation of the WIZARD franchisor's *website.*
Organization: Silmara Bernardino da Silva, 2016 (Prepared using *PHILCARTO software).*

Table 8 - Distribution of WIZARD units

UF	№ UNITS	UF	№ UNITS	UF	№ UNITS	UF	№ UNITS
CA	2	GO	34	PE	19	SP	389
AL	6	MA	12	PI	6	SE	8
AP	4	MS	12	RJ	111	TO	3
AM	10	MT	24	RN	11		
BA	29	MG	118	RS	112		
EC	10	PA	11	RO	5		
DF	39	PB	7	RR	1		
ES	21	PR	99	SC	78		

Source: Consult the franchisor's website.
Org.: Silmara Bernardino da Silva (2016).

From all the information presented, the concentration of units in the state of São Paulo and in the southeast and south regions, respectively, is significant. This characteristic is reinforced when we take the hierarchical classification that cities present based on the multiple and distinct urban activities available on their territory and as a reflection they can be classified as metropolises, medium-sized cities or local cities according to their degree of diversification. Making up the urban network, which is the articulated set of urban centers that in the past were hierarchical in structure due to the low functional complexity of the urban centers, a few had a greater degree of specialization and offer of goods and services, making them perform a regional function when taking the territorial division of labor, which was quite simplistic at the time.

The globalization phenomenon has had repercussions on the Brazilian urban network, inserting it into dynamics both internal and external to the network itself, reshaping it in terms of size, density, functions, nature, intensity and spatial scope of interactions and its form. This phenomenon causes cities to be organized in a dispersed and concentrated manner, leading to more articulated and diffuse inter-scalar interactions (Correa, 2001; Catelan, 2013).

The complexification drawn from the new ramifications of the urban network reveals the same process that materializes in the urban space from the new functions and spatial patterns leading them to articulate networks on multiple scales.

Within the urban network, the functions and roles of cities are defined in a back and forth of logics, interests and dynamics intrinsic to the reproduction of capital, whose movement comes from the process of articulating scales. They generate inter-scalar spatial interactions that explode in the hierarchical structure of the network, with points appearing in space that show greater complexity than if thought of at their hierarchical level. These are scales, especially the regional one, which not only survive, but become essential in the face of movements such as the advance of global economic agents. The network is becoming more complex, times are overlapping, spaces are (re)signifying themselves and spatial interactions are densifying (CATELAN, 2013.p.67).

Reflecting on the author's proposition, we can relate it directly to our object and the city of study, pointing out some pertinent observations on the subject of interactions, the changing role played by the city and the agents involved. To carry out the exercise of relating theory to our empirical observations, let's first look at Table 8.

Table 9 - Origin of franchising companies

FRANCHISE	LOCATION OF HEADQUARTERS	SCALE OF OPERATION	№ BRAZIL UNITS	OUTDOOR UNITS
CNA	Sao Paulo - SP	National	602	
FISK	Sao Paulo-SP	International	1000	Angola -2/ Argentina - 82 /Japan - 15/Paraguay - 2 / Bolivia - 1 / Chile 1.
BRAZILIAN LANGUAGE INSTITUTE - IBL	Maceió - AL	Except Northern Region	33	
KUMON	Japan (worldwide) New Jersey (USA) ' Sao Paulo - (South American) Presidente Prudente	International Regional	1427	Europe - Andorra, Austria, Belgium, France, Germany, Greece, Hungary, Ireland, Italy, Luxembourg, Netherlands, Spain, Switzerland, U.K. / Africa - Botswana, Kenya, Namibia, South Africa / Middle East - Bahrain, Qatar / Oceania - Australia, New Zealand / Asia - Brunei, China, India, Indonesia, Japan, Korea, Malaysia, Myanmar, Philippines, Singapore, Sri Lanka, Thailand, Vietnam, Hong Kong, Macau / North America - Canada, USA and Mexico. South America - Argentina, Brazil, Bolivia, Chile, Colombia, Peru.
MAPLE BEAR	Vancouver- Canada	International	103	Bangladesh, Singapore, India / Brazil South Korea / Morocco / Canada / United Arab Emirates / Mexico / China / USA / Vietnam / Total-132.
MICROCAMP (Holding - ABC, school, MC Viagem eTurismo, A Fazenda MC, Red Anqus restaurant	Campiñas - SP	National / International	80	
MICROLINS	Sao José do Rio Preto - SP	National	411	
OPTIMUM ENGLISH PROFESSIONAL	Rio de Janeiro - RJ	Regional - RJ/SP/GO	8	
WIZARD (Multi Holding Group - Pearson Group)	Campiñas -SP	International	1.212	USA-2 Costa Rica - 1/ Paraguay - 3 / Japan - 3
YÁZIGI (Holding- Pearson Group)	Sao Paulo - SP	National	277	

Information source: institutions' websites - July /2016

The table shows information on the location of the headquarters of the franchising companies in Brazil for the Brazil for the brands surveyed in Presidente Prudente.

The headquarters are distributed mainly in the Southeast, Sao Paulo, Campinas and Rio de Janeiro, while only one of them is located in the northeast of the country, Maceió, i.e. these cities represent the administrative, commercial and management centers of the brands. This in itself shows us how the city under study interacts with other urban spaces and even international urban realities, since its headquarters are located in other countries. In order to reinforce and illustrate the processes already highlighted, we should now consider the distribution and classification of the Fisk, Kumon and Wizard franchises[6] based on the Regions of Influence of Cities (REGIC), carried out by the Brazilian Institute of Geography and Statistics (IBGE). Table 10 shows the typology used to classify urban centers.

It is important to note that at the top of the urban network we have the metropolises, as they concentrate the most important roles in the management of the territory, both in terms of political and economic power. In these areas, the importance of economic agents is clear, especially those linked to large corporations, whose actions start to order the spread of capital. Immediately below the metropolises are the regional capitals, which have a direct relationship with the metropolises and polarize the region around them. Regional Capitals C have an average of 250,000 inhabitants. It is in this category that the city studied is classified. The information on the classification of regional capitals is in line with the perception of the city constructed by the franchisee of the Brazilian Language Institute, who in his study found that Presidente Prudente is a regional hub. The hierarchical classifications below the regional capitals are Sub-regional Centers A and B, made up of 169 centers with less complex management activities, which have a smaller scope of action, and their relationships with external centers are limited to three national metropolises. Located immediately below the sub-regional centers are the zone centers, which can be characterized by 556 smaller municipalities with operations restricted to their immediate area and management functions elementary.

[6] Some cities were not included in the franchise count because they were not listed in the official REGIC publication or in its database. See Appendix 12 for a list of the cities.

Table 10 - Classification of the Brazilian Urban according to REGIC, 2007.

Classification	Main Characteristics
Metropolis	Made up of 12 cities, they are characterized by their large size and strong ties with each other, as well as generally having an extensive area of direct influence.
Large Metropole Naaonal	Made up of 1 city, it represents the first level of territorial management
National Metropole	Made up of 2 cities, these centers are at the first level of territorial management and will be the focus for focused centers throughout the country.
Metropole	Made up of 9 cities, this is the second level of territorial management
Regional Capital	Made up of 70 cities, like the metropolises, it is also related to the upper stratum of the urban network. With management capacity at the level immediately below that of the metropolises, it has a regional area of influence and is referred to as a destination for a range of activities by a large number of cities.
Regional Capital A	Consisting of 11 cities, with an average of 955 million inhabitants and 487 connections.
Regional Capital B	Made up of 20 cities, with an average of 435,000 inhabitants and 406 connections.
Regional Capital C	Consisting of 39 cities with an average of 250,000 inhabitants and 162 relationships.
Sub-regional center	This level includes 169 centers with less complex management activities, predominantly between levels 4 and 5 of territorial management, they have a smaller area of operation, and their relations with centers outside their own network are generally only with the three national metropolises.
Sub-regional Center A	Made up of 85 cities, with an average of 95,000 inhabitants and 112 links
Sub-regional Center B	Consisting of 79 cities, with a median of 71,000 inhabitants and 71 connections
Zone Center	Made up of 556 smaller towns with operations restricted to their immediate area; they perform elementary management functions.
Zone Center A	Made up of 192 cities, with a median population of 45,000 and 49 predominant territorial management levels 5 and 6 (94 and 72 cities, respectively), with nine cities at the fourth level and 16 not classified as management centers.

Zone B Center	Made up of 364 towns, with an average of 23,000 inhabitants and 16 settlements. The majority, 235, had not been classified as a territorial management center, and another 107 were at the last level of that classification.
Local Center	The other 4,473 towns whose centrality and activity do not extend beyond the boundaries of their municipality, serving only their inhabitants, have a dominant population of less than 10,000 inhabitants (median of 8,133).

Source: REGIC, IBGE (2008).
Org.: Letícia M. Sant' Anna (2014).

After gathering some explanatory elements about the methodology used by REGIC, we'll start our observations with Graph 2, which shows the distribution of FISK, KUMON and WIZARD franchise units across the country's regions.

Graph 2 - Relation of franchised units in Metropolis.

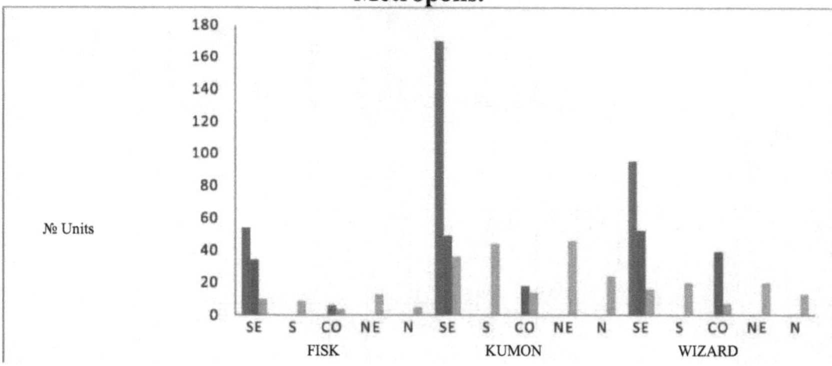

	SE	S	CO	NE	N	SE	S	CO	NE	N	SE	S	CO	NE	N
■Major National Metropolis	54	0	0	0	0	170	0	0	0	0	95	0	0	0	0
■National metropolis	34	0	6	0	0	49	0	18	0	0	52	0	39	0	0
■Metropolis	10	9	4	13	5	36	44	14	46	24	16	20	7	20	13

The blue column in the graph represents the major national metropolis, Sao Paulo, which is classified at the first level of territorial management. There is also a predominance of the Southeast as the region concentrating language franchises; in descending order, we have Kumon, Wizard and Fisk. Perhaps one fact we noticed during the interview and even when consulting the company's *website* was that Sao Paulo operates as the national and international headquarters, representing the brand's management center for Latin America, as shown in Chart 8.

Graph 3 shows the distribution of franchises in the Regional Capitals. At this point, it should be noted that the city we studied is classified as Regional Capital C.

Graph 3 - List of franchised units in Regional Capitals.

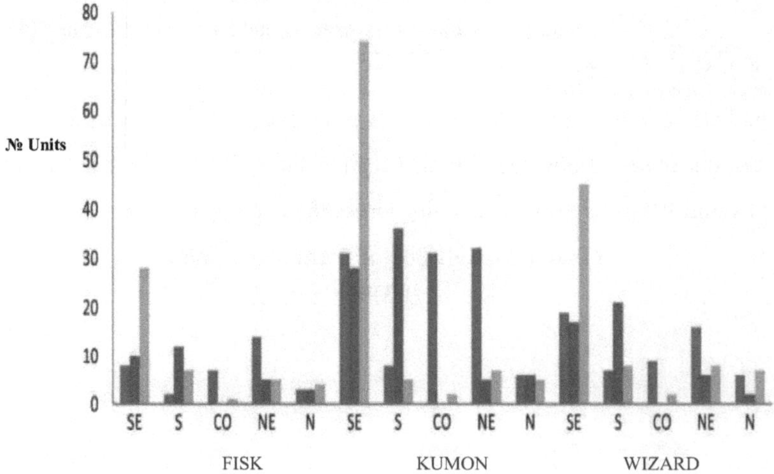

■Regional Capital A	8	2	7	14	3	31	8	31	32	6	19	7	9	16	6
■Regional Capital B	10	12	0	5	3	28	36	0	5	6	17	21	0	6	2
■Regional Capital C	28	7	1	5	4	74	5	2	7	5	45	8	2	8	7

The cities classified in this range have a management capacity below that of the metropolises, their area of influence is regional and they are cited as the reference point for many activities carried out by a large number of municipalities. We can see from the graph that for the southeast region, Kumon again stands out as having the highest concentration of units, and even the other two brands stand out in the graph for having the highest values for the C Regional Capitals of the Southeast Region.

Graph 4 - Ratio of franchised units in Sub-Regional Centers.

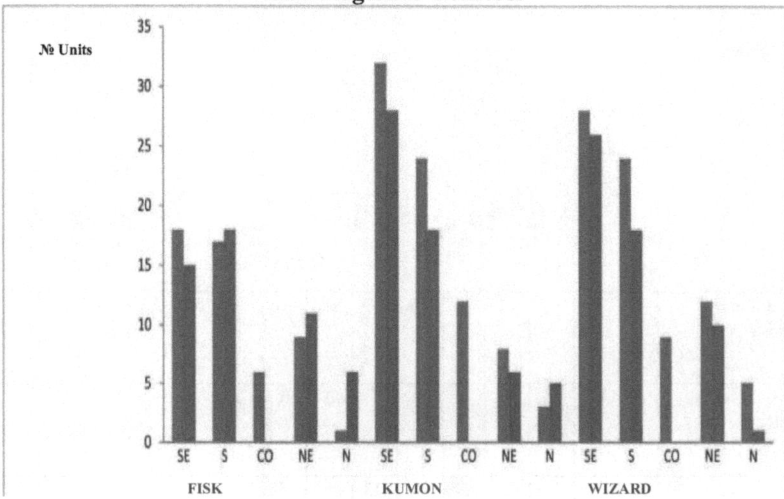

▪Sub-regional Center A	18	17	6	9	1	32	24	12	8	3	28	24	9	12	5
▪Sub-regional Center B	15	18	0	11	6	28	18	0	6	5	26	18	0	10	1

Sub-regional centers A and B, identified by Graph 4, are considered to carry out less complex management activities, and their relationship with other cities is reduced and the degree of interaction occurs with a few metropolises. These centers have populations of between 71,000 and 95,000 inhabitants, which explains their slow interaction with the urban network. There is a greater balance between the distribution of the Kumon and Fisk brands in the Southeast, but it is in these urban areas that the franchises are more dispersed when we look at the bars for the South and Northeast.

List of franchised units in Zone and Local Centers.

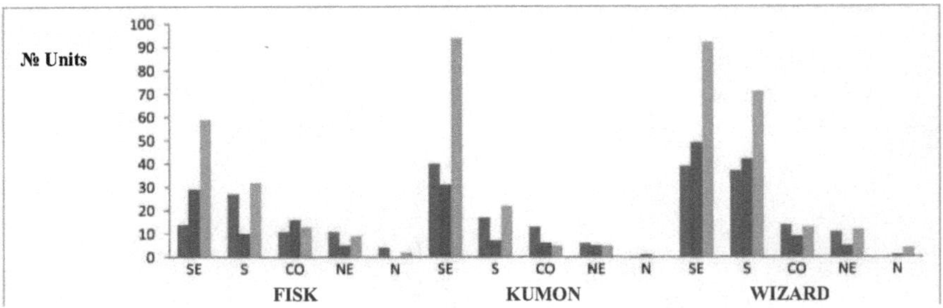

■ Zone Center A	14	27	11	11	4	40	17	13	6	0	39	37	14	11	0
■ Zone B Center	29	10	16	5	0	31	7	6	5	1	49	42	9	5	1
■ Local Center	59	32	13	9	2	94	22	5	5	0	92	71	13	12	4

Graph 3 contains the smallest centers, small in size and limited to their area of concentration, with very simple management functions. The pattern observed in the other graphs is repeated, with the Southeast always showing the highest number of units.

Denser reflections are still possible in this work, but to do so would require us to look at the methodology and materials used in this larger study and then articulate them would probably result in elements for the construction of a second work. Thus, by quantifying the spatialization of franchises by region and hierarchical level, we aim to demonstrate the extent to which this commercial system has expanded its fronts of action and, by locating itself in different parts of the territories, it hints at the multiple possibilities of articulation between scales, providing the elements that allow us to identify the overcoming of the predominance of the classic hierarchical pattern between cities.

FINAL CONSIDERATIONS

The purpose of this study was to investigate how franchised language schools, based on their logic of location in the city, build internal relationships with the city as a whole, taking into account the relationships between the cities in their urban network as well as with cities in other city networks. And to see if these relationships add up to the set of transformations through which the urban space has been significantly restructured, recognized by urban theorists as urban restructuring.

The objectives developed, as well as the methodology, were aimed at obtaining information and data that would allow us to point out the processes and theoretical concepts related to the new shape of the city.

We therefore believe that the objectives of identifying the franchised schools operating in the city and the dynamics that guide their different locations allowed us to discover elements that converge with urban theory as a whole, which points out that medium-sized cities, and even others of different sizes, have gained prominence as a result of the new spatial logics engendered by economic agents. The number of brands and the number of language franchises reinforce these processes and point out how consumption has been fundamental in guiding these actions.

It was possible to see the coexistence of many processes, including centralization, i.e. even though franchised schools are able to choose new locations in the city, there are a considerable number of them in the central area, sub-centres and main axes, reinforcing existing centralities and also giving the city potential for the development of new centralities.

The study of the specificities of these commercial operating systems, the franchise system, contributes to our understanding of the complexification of relationships in and between the urban space and urban networks, since we found that the franchising headquarters of the brands present in the city are located in other states of the southeast region and even in another region, in addition to the international origin and operation of some brands, which illustrates the face of globalization influencing urban practices and production.

The survey of the number of schools franchised in the cities and their classification based on the typologies of the urban centers reinforces the propositions of inter-scalar interactions, since we found that in Presidente Prudente there is a branch of an international franchise that polarizes and coordinates dozens of units located in cities around it and even in municipalities located in another state of the federation.

The systematization of the reflections and information gathered and produced about the city and the franchised and non-franchised language schools, added to the effort made to explain the theoretical understanding that forms the basis of this book, has led us to realize that progress has been made in the course of the study. These advances are mainly linked to the maturing and understanding of how to carry out research and the role and importance that geography plays. It is up to the geographer to unravel the multiple processes, be it the dimension of the inhabited space or the world-space, and for this journey there are many geographical lenses that can and should complement each other. We hope that this text, with all its limitations, can contribute in some way to new research and studies on the multiple dynamics of medium-sized cities.

BIBLIOGRAPHICAL REFERENCES

ANTAS JR. R. NOTE ON THE USE OF THE CONCEPT OF CIRCUITS SPACE PRODUCTIVE TO ESTABLISHING THE NEXUS BETWEEN URBAN RESTRUCTURING AND THE REFUNCTIONALIZATIONS OF SPACE: A STUDY ON HEALTH FACILITIES IN THE STATE OF SÁO PAULO. In: Encontro Nacional da ANPUR, XIX, Rio de Janeiro. p.3-14.

BAENINGER, R. Urban restructuring: some considerations on the current debate. 1998. **Regions and Cities on the Tiete-Paraná Waterway.** Campiñas: Center for Population Studies, UNICAMP, 2004. 101p.

BARBOSA, L. A. **The consumer society.** Rio de Janeiro: Jorge Zahar, 2004. 68 p.

CARLOS, A. F. A. **The City.** 9. ed. Sao Paulo: Contexto, 2011.

CARLOS, A. F. A; SOUZA, M. L; SPOSITO, M. E. B (org.). **THE PRODUCTION OF THE URBAN SPACE**: agents, processes, scales and challenges. 1. ed, 3ªreprint. Sao Paulo: Contexto, 2014.

CATELAN, M. J. **Heterarquia urbana:** interapóes espaciais, interescalares e cidades medias. Sao Paulo: Cultura Académica, 2013. ePDF.

CHERTO, M. R. **FRANCHIISNG:** the marketing revolution. Sao Paulo: McGraw-Hill, 1988.

CORRÉA, R.L. **O Espapo Urbano.** Sao Paulo, Editora Ática, Série Princípios, 3rd ed, n. 174.

. The Brazilian urban network and its dynamics: some reflections and questions. In: **URBANIZATION AND CITIES:** GEOGRAPHICAL PERSPECTIVES. Sposito (org). Presidente Prudente, GAsPERR, 2001. 643p.

. Processes, forms and spatial interactions. In: **Revista Brasileira de Geografia**, Rio de Janeiro, v. 61, n.1, p.127-134, 2016.

LEFEBRVE, H. **The right to the city.** 5 eds., 5ªreimp. Sao Paulo: Centauro, 2001.

LIMONAD, E. Reflexoes sobre o Espaço, o Urbano e a Urbanizaçao. **Revista GEOgraphia,** Rio de Janeiro, v. 1, n. 01, p. 71-91, 1999.

PRESIDENTE PRUDENTE. Complementary Law No. 153, of January 10, 2008. Florivaldo Leal Municipal Palace.

ORTIGOZA, S. A. G. **FRANCHISES AND THE NEW STRATEGIES OF URBAN COMMERCE IN BRAZIL**. 1996. 180 p. Master's dissertation in Geography -
Institute of Geosciences and Exact Sciences, Paulista State University, Rio Claro.

PORTO-SALES, A. L. **THE SPACE SITUATION OF FRANCHISES IN SOUTH AMERICA**: urban morphology and centrality in medium-sized cities in Argentina, Brazil and Chile. 2014. 250 p. PhD Thesis in Geography - Faculty of Science and Technology Universidade Estadual Paulista, Presidente Prudente.

Regions of Influence of Cities 2007- **REGIC**. Rio de Janeiro: IBGE, 2008.

SANTOS, M. O PAPEL ATIVO DA GEOGRAFIA UM MANIFESTO. In: Encontro Nacional de Geógrafos, XII, 2000, Florianópolis. p. 103-109.

__________________. **Technique, Space and Time:** Globalization and the Technical-Scientific-Informational Environment. 5th ed., 1. Reimp.- Sao Paulo: Editora da Universidade de Sao Paulo, 2013. 176 p.

SINGER, P. The use of urban land in the capitalist economy. In: MARICATO, Ermínio, org. **A produção capitalista da casa (e da cidade) no Brasil industrial**. Sao Paulo, Alfa- Omega, 1978. p 21-36.

FAPESP RESEARCH. The economic relief of the interior. Sao Paulo: FAPESP, Ed. 197, Jun, 2012. 98 p.

SOJA, E. W. **Postmodern Geographies:** the reaffirmation of space in social theory. Translation [of the 2nd English edition], Rio de Janeiro: Jorge Zahar, 1993. 323 p.

SPOSITO, M. E. B. As cidades médias e os contextos econômicos contemporâneos. In: Sposito, M. E. B. [org.] **Urbanization and cities:** geographical perspectives. Presidente Prudente: GASPERR, 2001, p. 609-643.

_________________________. Medium-sized cities: city restructuring and urban restructuring. In: SPOSITO, M. E. B. (org.). **Cidades médias espaços em transiçao**. Sao Paulo: Editora Expressao Popular, 2007, p. 609-643.

_________________________. **Economic logics and contemporary spatial practices:** medium-sized cities and consumption. Presidente Prudente: Project submitted to FAPESP, 2011.

________________________. New commercial forms and the redefinition of interurban centrality. In: Sposito, M. E. B. (org.) **Textos e Contextos para a Leitura de uma Cidade Média.** Presidente Prudente: GASPERR, 2001. p.235-253.

__________________________. The Center and the forms of expression of centrality. **Revista Geografia,** Sao Paulo, v. 10, p. 118. 1991.

____________________________. Production of Space and New Urban Territorialities in the State of Sao Paulo. In: VIII Meeting of Latin American Geographers, 2001, Santiago. Anales: Santiago, Universidad de Chile, Facultad de Arquitectura y Urbanismo, 2001. p. 187-197.

____________________________ Socio-spatial Segregation and Urban Centrality. In: VASCONCELOS, P. A; CORRÊA, R. L; PINTAUDI, S. M. **A cidade contemporânea**: segregaçao sociespacial. Sao Paulo: Contexto, 2013. P. 6193.

. Diffuse urbanization and dispersed cities: perspectives perspectives. Sao Paulo: Via das Artes / FAUUSP, 2009, p.38-54.

VILLAQA, F. **Intra-urban space in Brazil.** Sao Paulo, Studio Nobel: FAPESP. Lincoln Institute, 2011. 363 p.

ZILHÁO, A. Uma proposta multidisciplinar para o entendimento da centralidade urbana como facto social total. **Revista da Faculdade de Letras,** Porto, v. XXV, p. 55-77, 2013.

WEBSITES CONSULTED

Brazilian Franchising Association. Provides information and data on the sector's segments, research, etc. Available at: http://www.abf.com.br.

Lista Mais. Information on establishments in the city with address and telephone number. Available at: http://www.listamais.com.br

FISK. Provides information about the school, its units, brand history, etc. Available at: http://www.fisk.com.br

WIZARD. Provides information about the school, its units, brand history, etc. Available at http://www.wizard.com.br

KUMON. Provides information about the school, its units, historyhistory etc. Available at http://www.kumon.com.br

MapleBear. Featuresinformation about the school, its units, history the etc. Available at http://www. maplebear. com. br/presidenteprudente

Calenglish. Presentsinformation information about the school, location, methodology, etc. Available at http://www.calenglish.com.br

Optimum Tech. Provides information about the school, its units, brand history, etc. Available at http://www.optimumtech.com.br

CNA. Provides information about the school, its units, brand history, etc. Available at http://www.cna.com.br

YÁZIGI. Provides information about the school, its units, brand history, etc. Available at http://www.yazigi.com.br

IBL. Provides information about the school, courses, its franchise system, its units, brand history, etc. Available at http://www.ibl-idiomas.com.br

Pequenas Empresas Grandes Negócios magazine. Provides news on the new entrepreneurs segment, the market for the sector, including the franchising sector, research, etc. Available at http://www.revistapegn.com.br

Rossi Comunicapóes. He presents data on the companies to which he provides communications consultancy, including several franchises. Available at http://www.rossicomunicacao.com.br

FSC
www.fsc.org
MIX
Papier | Fördert gute Waldnutzung
FSC® C083411

Kolibri 360 GmbH
Ferdinand-Jühlke-Straße 7
99095 Erfurt, Deutschland
produktsicherheit@kolibri360.de

Druck:
CPI Druckdienstleistungen GmbH
im Auftrag der
Kolibri 360 GmbH
Ferdinand-Jühlke-Str. 7
99095 Erfurt